THE U.S. NAVY SEAL SURVIVAL HANDBOOK

LEARN THE SURVIVAL TECHNIQUES AND STRATEGIES OF AMERICA'S ELITE WARRIORS

Don Mann
and Ralph Pezzullo

Skyhorse Publishing

Skyhorse Publishing books may be purchased in bulk at special discounts for sales promotion, corporate gifts, fund-raising, or educational purposes. Special editions can also be created to specifications. For details, contact the Special Sales Department, Skyhorse Publishing, 307 West 36th Street, 11th Floor, New York, NY 10018 or info@skyhorsepublishing.com.

Skyhorse® and Skyhorse Publishing® are registered trademarks of Skyhorse Publishing, Inc.®, a Delaware corporation.

Visit our website at www.skyhorsepublishing.com.

10

Library of Congress Cataloging-in-Publication Data is available on file.
ISBN: 978-1-61608-580-3

Photo research by Erika W. Hokanson of RefreshMediaResources.com

Printed in China

"Never Get Lost AGAIN! As a former Navy SEAL, not only did I enjoy Don Mann's *U.S. Navy SEAL Survival Handbook*, but learned several things as well. This book is written by a career Navy SEAL combat medic and life long adventure racer. He knows more about surviving in potentially dangerous situations/nature than anyone I personally know. I look forward to keeping this book readily available—just in case!"

—by Stew Smith, Navy SEAL Fitness Author

"Don has certainly walked the walk in his years serving in the SEAL teams. Having gone through military survival training (land/water survival, SAR, SERE, etc.) and non-military (winter mountaineering) I can appreciate the depth of knowledge within each section. Surviving life threatening situations requires something unique for each individual. Don articulates a holistic view to help educate and prepare the reader for various scenarios. This publication is a must read for anyone looking to hone their survival skills."

—by Brian Dickinson, ex-USN CSAR Swimmer, Everest Summiteer/Survivor

"*The Navy SEAL Survival Handbook* is a comprehensive guide from an expert with the experience to back it up! *The Navy Seal Survival Handbook* is a terrific, one stop source for your survival needs."

**—by Randy Spivey, CEO and Founder,
Center for Personal Protection and Safety**

"I've been on multiple operational deployments with Don and taken a few recreational adventure trips with him. I assure you that Don knows about what he has written in *The Navy SEAL Survival Handbook*. I don't mean theory or practicing survival techniques in protected environment where a time-out can be called and someone will come and provide a little assistance. Don has applied most of the things he describes in the real-world and in risky situations. Heck, he's probably done all of this stuff but couldn't share it because of the sensitive nature of his missions! While Don describes skills that are critical to survival, what you really gain from reading this handbook is a sense of the attitude required to be successful in the harsh environments in which America's secret warriors thrive."

**—by John Wright, US Air Force and DoD SERE Trainer,
Multiple Deployments in Support of the Global War on Terror**

"*The Navy SEAL Survival Handbook* is essential reading for any adventure racer, climber, explorer, military personnel or outdoors person; ensure you have a copy of this in your kit-bag! Everything from what to include and how to prepare your emergency kit to living off the land, from essential shelter building to personal administration and from the effects of weather to celestial and obscure navigation can be found in this handbook.

I know this is one manual I will be using for the Hellweek PT™ Baalsrud Trail Challenge to the Arctic in 2013."

—by (Naval) Lt. Seán Ó Cearrúlláin (Rtd.), Hellweek PT™ Founder

"*The Navy SEAL Survival Handbook* has all the lessons, reasons why and also the why not's, of how to survive when "it" hits the fan. Everyone from Back Country to the Urban folks can benefit from all the information. You just never know and like it has been said "knowledge is power". The lessons in this book can mean the difference between surviving and being home with those you love or not. Will you be part of the solution or live a life time of regret? I started reading it, but wound up studying it!"

—by Jim Kelleher, Survivalist

"*The Navy SEAL Survival Handbook* brought back some really wild and vivid memories of courses I'd been through in the military. It really captured what it was like to be there and how each evolution unfolded. Special Operation Forces (SOF) guys are pretty special just as the studies show. This Survival Handbook tells the public exactly how they are able to do survive in the wilderness."

—by LTC Blain Reeves, US Army

"Having been trained by some of the most highly regarded survival experts in the US, I can say definitively that no one has more experience or knowledge on the topic of mountain, jungle, desert, sea or urban survival than Don Mann. Don's survival techniques along with his philosophy on combat mindset have served me well in real world survival situations and life in general. *The Navy SEAL Survival Handbook* should be mandatory reading for anyone who spends time in the wilderness."

—by Dr. Stephen M. Erle, Director SEAL Training Adventures

"A veritable how-to guide, *The Navy Seal Survival Handbook* is an absolute must have for every outdoor enthusiast, adventurer or Armed Forces member. This book is packed with a ton of useful information on wilderness survival techniques. A retired member of the elite U.S. Navy SEALs, Don Mann shares his acumen and experiences as an expert in the field."

—by Rick Sheremeta, Outdoor Photographer and Author

"Navy SEAL, Don Mann, comes through again with another insightful and inspiring book. His knowledge of survival is apparent from the get go in *The Navy SEAL Survival Handbook*. The psychological angle intrigued me the most and was an amazing view of what SEALs sacrifice IN TRAINING, let alone what happens in combat situations. I gained a great deal of insight into the world of survival as well as deepened my respect for the Navy SEALs. Frankly if I was in a survival situation, I would want Don Mann there with me. After reading his book I at least feel like I am much better prepared. If you are a fan of survival, get this book. If you are intrigued by the Navy SEALs, get this book. Just GET THIS BOOK!"

—by Matthew Fox, Expedition competitor

"I found the chapters on Survival Mindset and Basic Survival Tips to be especially beneficial. These two chapters alone are a vital read for all outdoors recreationalists. I highly endorse this book as the definitive guide on wilderness survival and recommend it strongly as essential reading for military, civilian, and search and rescue specialists."

—by Jim C. Blount, retired CIA Senior Instructor
and the author of *Stay Alert, Stay Alive, A Guide to Counterterrorism for Everyday Life*

"*The Navy SEAL Survival Handbook* is much more than a basic "Survival 101" handbook. Having retired from the Army, after a twenty-four year career, I read from a perspective of does this handbook have applicability in my life today, as a business executive? Answer: YES! It raises the business leader's awareness that we travel in a dangerous world, that knowledge, skill and proper preparation is required to successfully meet our daily challenges, a "can do" attitude positively affects those around us and mental toughness is a required trait for success. This is a great read, for the special ops soldier, survivalist, adventurist, and business executive!"

—by Terry L. Carrico, VP, Corporate Security,
McKesson Corp. Colonel (R), U.S. Army, Military Police Corps

CONTENTS

INTRODUCTION

"It is not the strongest of the species
that survive, nor the most intelligent,
but the one most responsive to change."
—*Charles Darwin*

Life can change in an instant. One moment you're trekking along a mountain ridge, and then you're suddenly struck by a rattlesnake. One minute you're skiing down a mountain, the next you're facing an oncoming avalanche.

As Navy SEALS, we are trained to face dangerous situations all over the world—at sea, in the air, and on land. We understand that every time we launch a mission, unforeseen circumstances—flash floods, aircraft crashes, poisonous insect, animal or marine life bites—can cause us to be trapped in an unfriendly environment, cut off from communication to outside support. With every tick of the clock, our situation can become more desperate. We are trained to deal with these emergencies. We learn how to survive in the harshest of environments without food and water.

U.S. service members participating in a Survival, Evasion, Resistance and Escape (SERE) course march to their first field training evolution at a training site in Warner Springs, Calif. Personnel attending the SERE course are trained in evading capture, survival skills and the military Code of Conduct.

But what if, God forbid, something like that happens to you? Will you know how to treat yourself if you're bitten by a poisonous snake? Will you know what plants you can extract water from if trapped in the desert? Or how to protect yourself and your teammates from freezing to death in the mountains? The answer to all these questions should be "yes."

This handbook is all about developing the SEAL survival mind-set, and arming yourself with the appropriate survival techniques for numerous potentially fatal scenarios.

We live in a dangerous world. It's your responsibility to be prepared.

—*Don Mann*

Rangely, Maine—A student at the Navy Survival, Evasion, Resistance and Escape (SERE) school crosses a frozen creek.

U.S. sailors undergoing the third and final phase of Basic Underwater Demolition/SEAL training use a rope to guide themselves down the side of a cliff and into the ocean during a field training exercise at San Clemente Island, Calif. The third phase of the training provides the students with skills in small arms weapons, demolitions, and tactics, which culminate in the planning and execution of various missions as an independent platoon.

1

SEAL/SERE
TRAINING

"Return with honor."
—*SERE school motto*

U.S. Navy SEALs

≈ **U.S. Navy SEAL candidates from Basic Underwater Demolition/SEAL (BUD/s) Class 288 participate in log physical training (log PT) during the first phase of training at Naval Amphibious Base Coronado, Calif.**

I've spent my adult life as a Navy SEAL, preparing for and dealing with the most dangerous situations imaginable.

From 1962, when the first SEAL teams were commissioned, to the present, SEALs have distinguished themselves as being individually reliable, collectively disciplined, and highly skilled. Because of the dangers inherent in what we do, prospective SEALs go through what is considered by military experts to be the toughest training in the world—Basic Underwater Demolition SEAL Training (BUD/S).

BUD/S is a six-month SEAL training course held at the Naval Special Warfare Training Center in Coronado, California, which starts with five weeks of indoctrination and pre-Training. Following that, all trainees go through three phases of BUD/S. The first phase is by far the toughest and consists of eight weeks of basic conditioning, with a grueling "hell week" in the middle—which is five days and nights of continuous training on a maximum of four hours of sleep.

Hell week is a test of physical endurance, mental tenacity, and teamwork. As many of two-thirds of the class are likely to "ring the bell" and call it quits during this phase. Those who grit it out to the finish get to hear the instructors yell, "hell week is secured!" The trainees continue on with a

Coronado, Calif. A student in Basic Underwater Demolition/SEAL (BUD/S) class 270 navigates his way through the chaos of smoke and explosions in one of the final evaluations of hell week. On average, students are allowed only four hours of sleep during hell week, and those who complete it have about a 95 percent chance of graduating BUD/S. ≈

Basic Underwater Demolition/ SEAL (BUDs) students participate in Surf Passage at Naval Amphibious Base Coronado. ≈

new sense of pride, achievement, and self-confidence to second phase (eight weeks of diving) and third phase (nine weeks of land warfare).

After BUD/S is completed, all trainees go through three weeks of basic parachute training, followed by eight weeks of SEAL qualification training in mission planning, operations, and tactic, techniques, and procedures.

BUD/S ends with the formal BUD/S class graduation. It was a very proud day for me to stand with my classmates in our dress navy uniforms and listen to our SEAL officers talk about the special group we were about to enter, and the great honor it is to serve as a U.S. Navy SEAL.

BUD/S Phases

Phase 1—Physical Conditioning (eight weeks)

Soft sand runs

Swimming—up to two miles with/fins in the ocean

Calisthenics

Timed obstacle course

Four-mile timed runs in boots

Small boat seamanship

Hydrographic surveys and creating charts

Hell week—week 4 of phase 1—five and one-half days of continuous training on little to no sleep

Phase 2—Diving (eight weeks)

Step up intensity of the physical training

Focus on combat diving

Open-circuit (compressed air) SCUBA

Closed-circuit (100% oxygen) SCUBA

Long-distance navigation dives

Mission-focused combat swimming and diving techniques

Phase 3—Land Warfare (nine weeks)

Increasingly strenuous physical training

Weapons training

Demolitions (military explosives)

Small unit tactics

Patrolling techniques

Rappelling and fast rope operations

Marksmanship

⌃ **As an instructor monitors a training evolution, Basic Underwater Demolition/ SEAL (BUDS) Class 244 receives instructions on their next exercise while they lay in the surf.**

Kodiak, Alaska—Navy SEALs perform advanced cold weather training to experience the physical stress of the environment and how their equipment will operate, or even sound, in adverse conditions. Navy SEALs are maritime special operations forces that strike from the sea, air, and land. They operate in small numbers, infiltrating their objective areas by fixed-wing aircraft, helicopters, navy surface ships, combatant craft and submarines. SEALs have the ability to conduct a variety of high-risk missions, utilizing unconventional warfare, direct action, special reconnaissance, combat search and rescue, diversionary attacks and precision strikes. ❯

All BUD/S graduates then fly out to Kodiak Island, Alaska, for a twenty-eight day winter warfare course, during which they train in snow and freezing wind while often carrying half their body weight in weapons and gear. The course includes cross-country skiing, snow shoe travel, building shelters, procuring food and water, fire building, using specialized survival gear to plot courses in the mountainous and snow-covered terrain, and conducting ice-cold ocean swims, river crossings, and long-range navigation through the mountain wilderness to infiltrate and establish covert surveillance of target sites.

BUD/S and winter warfare training prepares SEAL trainees to become combat-ready warriors. But they don't learn the nitty gritty of survival until they complete SERE School.

Kodiak, Alaska—A SEAL qualification training candidate looks out from a two-man tent during a re-warming exercise in which he spent five minutes in near freezing water. ❯

« Kodiak, Alaska—A SEAL qualification training candidate checks the gear of another member of his squad during a long-range land navigation exercise. The candidates will spend forty-eight hours in the Alaskan mountains learning how to navigate through the rugged terrain and survive the frigid conditions. The twenty-eight-day cold weather training course, taught in Kodiak, is part of a year long process to become a U.S. Navy SEAL.

SERE Training

As a young Navy SEAL recently graduated from Basic Underwater Demolition School (BUD/S), I was told by a Vietnam-era SEAL that if I were captured during wartime, there was a good chance I'd be beheaded or skinned alive. I immediately volunteered to attend the Navy Survival Evasion Resistance and Escape (SERE) course conducted at Warner Springs, California.

Since I knew that as a SEAL I would likely be deployed overseas behind enemy lines, I took my survival training seriously.

Most of the twenty members in my SERE class were navy pilots and aircrew personnel considered to be at high risk of capture. I was the only SEAL.

The course started with basic lessons in land navigation, poisonous plants, animals and insects, water procurement, fire making, shelter building, and evasion and escape techniques. Then, the twenty of us were dropped off in the desert without food or water and ordered to find our way to a safe area while trying to avoid contact with the "enemy."

We were thirsty and hungry. We drank from the prickly pear cactus and looked for edible plants to eat. I happened to see a small rabbit running under a bush, threw my KA-BAR knife at it, and to my surprise, pinned the rabbit's neck into the ground. I skinned it and made rabbit stew for the team by mixing the rabbit with edible plants. But one little rabbit was hardly enough to feed twenty hungry men.

Eventually, all of us were captured. I was plastic-tie tied, blindfolded, and thrown into a Jeep. The instructors, outfitted in realistic communist-style clothing, played their parts, screaming, barking orders, trying their best to intimidate us.

I played it for real, too. When my captor stepped out of the Jeep, I managed to wrestle my bound hands in front of me, grab his PRC-77 radio, and throw it under the vehicle. I also hid a knife and lighter in my boots.

I was driven to a fenced POW training camp. There I saw enemy guards interrogating other "prisoners," slamming them into walls, humiliating them by having them stand naked while being drilled with questions and slapped in the face.

They started working on us immediately, trying to get us to break. There were hard cell interrogations with

U.S. Air Force Airmen of the 336th Training Group head out for a six-mile road march while wearing sixty-pound packs during the first ever Survival, Evasion, Resistance and Escape (SERE) Challenge held at Fairchild Air Force Base, Wash. Thirty SERE members are participating in the challenge, which includes an obstacle course, three-mile run, six-mile road march, and a variety of other events that test the strength and stamina of the participants. ≫

guards shouting questions and slapping you, and soft cell sessions, where you were called into a warm office where a pretty woman or friendly guard would offer you coffee, snacks, and warm clothing.

What is SERE Training?

Because of the violent nature of the world we live in, all U.S. military and other government personnel, and even civilians traveling overseas, run the risk of kidnapping, captivity, and exploitation by governmental and non-governmental organizations (including terrorist groups) that ignore the Geneva Convention and/or other human rights conventions. If you travel overseas frequently, especially to areas of conflict, you need to know how to avoid danger, evade capture, and, if captured, survive while waiting for extraction.

NATO countries provide basic level Survival Evasion Resistance and Escape (SERE) training to all their deployable forces.

SERE Training Levels

Level A is initial entry-level training that all military personnel—enlisted and officers—receive upon entering the service. It provides a minimum level of understanding of the code of conduct.

« **Getting the attention of four A-10 Thunderbolt II pilots by using a mirror during Combat Search And Rescue (CSAR) training near Osan Air Base, South Korea. Annual CSAR training is conducted by Survival Evasion Resistance Escape (SERE) instructors and is designed to reacquaint aircrew members with CSAR procedures and techniques.**

A soldier waits for an HH-60 Pave Hawk helicopter to extract him from a » simulated hostile area during a Survive, Evade, Resist and Escape (SERE) exercise in support of Red Flag–Alaska (RF-A) near Eielson Air Force Base, Alaska. Feeman is role-playing a downed pilot who was shot down behind enemy lines. RF-A is a Pacific Air Forces-directed field training exercise for U.S. and coalition forces flown under simulated air-combat conditions.

« As seen under night vision photography, U.S. sailors assigned to Basic Underwater Demolition/ SEAL class 281 carry an inflatable boat toward the surf during a first phase navigation training exercise in San Diego, Calif. First phase is an eight-week course that trains, prepares, and selects SEAL candidates based on physical conditioning, water competency; mental tenacity and teamwork.

Level B is designed for personnel whose "jobs, specialties or assignments entail moderate risk of capture and exploitation." Department of Defense Policy No. 1300.21 lists as examples "members of ground combat units, security forces for high threat targets and anyone in the immediate vicinity of the forward edge of the battle area or the forward line of troops."

Current operations in Iraq and Afghanistan have shown that practically everyone deployed in theater falls under this category. Consequently, the demand for Level-B training has proliferated exponentially and has become mandatory for most deploying forces. Level B is conducted at the unit level through the use of training/support packets containing a series of standardized lesson plans and videos.

Level C is designed for personnel whose "jobs, specialties or assignments entail a significant or high risk of capture and exploitation." According to military directive AR 350-30, "As a minimum, the following categories of personnel shall receive formal Level-C training at least once in their careers: combat aircrews, special operations forces (e.g., navy special warfare combat swimmers and special boat units—i.e., SEALs, Army Special Forces and Rangers, Marine Corps Force Reconnaissance units, Air Force special tactics teams, and psychological operations units) and military attaché."

SERE C-Level Training

The course spans three weeks with three phases of instruction.

Survival, Evasion, Resistance and Escape (SERE) students aboard a life raft prepare to perform a twenty-man egress during water survival training at Langley Air Force Base, Va. The training, conducted by SERE specialists, teaches students how to survive if they egress over water. The course also covers proper signaling, desalinating water, finding sustenance, and how to properly release from a parachute drag. ≈

« Students in the Survival, Evasion, Resistance and Escape (SERE) carry partners to obtain a 1,000-meter pace count during land navigation in the survival portion course at a training site in Warner Springs, Calif. The SERE course provided training in evading capture, survival skills and the Code of the U.S. Fighting Force.

Phase One consists of approximately ten days of academic instruction on the code of conduct and in SERE techniques that incorporate both classroom learning and hands-on field craft.

Phase Two is a five-day field training exercise in which the students practice their survival and evasion skills by procuring food and water, constructing fires and shelters, and evading tracker dogs and aggressor forces for long distances.

Phase Three takes place in the resistance-training laboratory, a mock prisoner-of-war camp, where students are tested on their individual and collective abilities to resist interrogation and exploitation and to properly apply the six articles of the code of conduct in a realistic captivity scenario. The course culminates with a day of debriefings in which the students receive individual and group feedback from the instructors. These critiques help students process everything they have been through to solidify the skills they applied properly and to correct areas that need adjustment.

SERE Training Objectives

Within SERE training, every student is taught to understand and practice techniques in the following procedures:

« Rangely, Maine—A student at the Navy Survival, Evasion, Resistance and Escape (SERE) school repacks his gear after a lesson.

Survival	Evasion	Resistance	Escape
• Ensure immediate survival. • Maintain psychological and physiological well-being. • Maintain normal body temperature. • Maintain adequate hydration. • Maintain sufficient caloric intake.	• Use of location aids. • Avoid detection while static and mobile. • Conduct evasion plan of action.	• Assess conditions of capture. • Protect sensitive information. • Mitigate physical and mental stress. • Maintain self and others (survive with dignity). • Limit the degree to which you are exploited. • Attempt to escape.	• Aid rescue. • Establish communications with friendly forces. • Carry out RV procedures. • Carry out extraction procedures.

U.S. Navy SERE training is conducted at the navy's remote training site in Warner Springs, California, and in the mountains of Bath, Maine.

Besides teaching survival, SERE is also an advanced code of conduct course. All military personnel get their initial code of conduct instruction during basic training, where they're taught an American service member's legal responsibilities regarding capture by enemy forces. But SERE training goes far beyond that. Because the school is a combination of courses designed for personnel with jobs that entail greater-than-normal risks of being stranded behind enemy lines or captured by enemy forces, students get a deeper insight into the philosophies behind the code.

Article I

* I am an American fighting in the forces which guard my country and our way of life. I am prepared to give my life in their defense.

Article II

* I will never surrender of my own free will. If in command, I will never surrender the members of my command while they still have the means to resist.

Article III

* If I am captured, I will continue to resist by all means available. I will make every effort to escape and aid others to escape. I will accept neither parole nor special favors from the enemy.

Article IV

* If I become a prisoner of war I will keep faith with my fellow prisoners. I will give no information or take part in any action which

might be harmful to my comrades. If I am senior, I will take command. If not, I will obey the lawful orders of those appointed over me, and will back them up in every way.

Article V

- When questioned, should I become a prisoner of war, I am required to give name, rank, service number and date of birth. I will evade answering further questions to the utmost of my ability. I will make no oral or written statements disloyal to my country and its allies or harmful to their cause.

Article VI

- I will never forget that I am an American, fighting for freedom, responsible for my actions, and dedicated to the principles which made my country free. I will trust in my God and in the United States of America.

The instruction starts with classroom work and then focuses on real-world applications of the code of conduct. Following the classroom part of the course, students begin to learn methods of avoiding capture by the enemy. Eventually, they are captured and enter resistance and escape training.

The SERE field instructors are highly motivated, well trained, and possess an immense knowledge of the subject. As instructors, they're part naturalist, part guide, part psychologist, and part mentor. Their expertise includes techniques for surviving in the arctic, desert, open ocean, jungle, and mountain regions, in combat and in captivity.

Much of the training at SERE School contains lessons learned by service members who made it back across enemy lines or spent time as prisoners of war. Their experience makes them highly valued advisors.

During the field phase of the course, students are introduced to specific methods of navigation through hostile territory. The rule regarding navigating is twofold. First, you need to figure out how to get where you're going without being spotted. Second, you have to reach specific locations on schedule. With the clock ticking, caution sometimes has to be sacrificed for speed, which can result in close calls.

Survival lessons are interspersed. These include: fire building, trapping, creating shelters, and finding edible plants.

The SERE School Experience

The following are the impressions of an army soldier who completed SERE training.

Warner Springs, Calif. U.S. Navy Survival, Evasion, Resistance and Escape (SERE) instructors teach members of Boy Scout Troop 806 of Coronado, Calif. how to trap food to survive in the wilderness. Sailors from the Naval Air Station North Island SERE detachment volunteered to train the Boy Scouts.

"Most of the modesty the students brought with them disappears very quickly. When they sleep, they huddle together to stay warm as the temperature dips into the twenties and frost coats their packs. When they're hiking, they know that everyone else is just as hungry and thirsty as they are. Not knowing what is coming next also bonds them. When surprises occur, they must act as a team. There is a chain of command for each group, as well as the entire class. The leaders are doubly challenged, as they are responsible for ensuring their team acts properly, no matter what comes up. When there are lapses in leadership and issues could have been avoided or resolved in the chain of command, the instructors take the group leaders aside later to advise them on appropriate responses."

In his book, *In the Company of Heroes*, retired 160th Special Operations Aviation Regiment pilot CW4 Mike Durant reflected on the SERE training he received at Camp Mackall in the winter of 1988 and the strength it gave him during his eleven-day captivity in Somalia in October 1993: "I came away (from SERE) with tools that I never believed I would ever really need, but even in those first seconds of capture at the crash site in Mogadishu, those lessons would come rushing back at me. Throughout my captivity, I would summon them nearly every hour . . . I thanked Nick Rowe [Colonel Rowe developed the rigorous Survival, Evasion, Resistance and Escape (SERE) training program] silently every day, for the lessons I learned in SERE training. I asked that God bless him, as I tried to plan my next move." (Nick Rowe died in 1989.)

Why Do Some Handle Stress Better Than Others?

Even though SERE School was a "theoretical setting," it taught me that some people are better at dealing with the stresses and strains of life than others. Why?

Dr. Andy Morgan of Yale Medical School set out to find a real-world laboratory where he could watch people under incredible stress in reasonably controlled conditions. He found one in southeastern North Carolina at Fort Bragg, home of the Army's elite Airborne and Special Forces. This is where the Army's renowned survival school (their version of Navy SERE school) is located. It's also where they practice something called stress inoculation. Based on the concept of vaccines, soldiers are exposed to pressure and suffering in training in order to build up their immunity. It's a form of psychological conditioning: the more shocks to your system, the more you're able to withstand.

While soldiers are frightened and worn down with sleep deprivation and lack of food, they're also interrogating them using enemy techniques used during World War II, Korea, and Vietnam. The sessions are known to be extremely tough.

For Morgan, POW school was the perfect place to study survival under acute stress. Even though the soldiers understood that they were in training, and, therefore, not in serious danger, Morgan's findings were revealing. During mock interrogations, prisoners' heart rates skyrocketed to more than 170 beats per minute for more than half an hour, even though they weren't engaged in any physical activity. Meanwhile, their bodies pumped more stress hormones than the amounts measured in aviators landing on aircraft carriers, troops awaiting ambushes in Vietnam, skydivers taking the plunge, or patients awaiting major surgery. The levels of stress hormones measured were sufficient to turn off the immune system and produce a catabolic state, in which the body starts to break down and feed on itself.

Morgan's research (which was the first of its kind) produced some fascinating findings in terms of what types of soldiers most successfully resisted the interrogators and stayed focused. Morgan examined two different groups going throughout this training: regular army troops and elite special forces soldiers, who are known to be especially "stress hardy" or cool under pressure. At the start, the two groups were essentially the same. But once the stress began, he saw significant differences. Specifically, the two groups released very different amounts of a chemical in the brain called neuropeptide Y (NPY). NPY is an amino acid produced by our bodies that helps regulate blood pressure, appetite, learning, and memory. It also works as a natural tranquilizer, controlling anxiety and buffering the effects of stress hormones like norepinephrine—also known as adrenaline. In essence, NPY is used by the brain to block alarm and fear responses and keep the frontal lobe working under stress.

Morgan found that special forces are superior survivors because they produce significantly greater levels of NPY compared with regular troops. And, twenty-four hours after completing survival training, NPY levels in Special Forces soldiers returned to normal, while those in regular soldiers remained significantly lower.

With much more NPY in their systems, special forces soldiers really are special. Not only are they able to remain clearheaded under interrogation stress because of their ability to produce massive amounts of natural antianxiety chemicals, but they also bounce back faster once the stress is removed. In the pressure of warfare, that's a major advantage.

Dr. Morgan's Findings at the Navy Diving and Salvage Training Center

At the elite Navy Diving and Salvage Training Center in Panama City, Florida, instructors have a quick way of figuring out who will be capable of accomplishing extremely dangerous underwater missions. They take young sailors (not SEALs), tie their hands behind their backs, bind their feet, strap a dive mask between their teeth (similar to BUD/S training), and then throw them into the pool. The challenge is to stay afloat and live.

"The more you struggle," Morgan said, "the harder it is to get air and the more tired you get. You just have to inhibit the powerful, incredible instinct to breathe and your anxiety and alarm."

Most trainees quickly realize that the only way to avoid drowning is to relax and sink to the bottom of the pool, kick off powerfully toward the surface, gasp for a little bit of air through clenched teeth, then fall back into the water and drop down to the bottom again.

During this testing, a large number of sailors black out. They simply don't get enough oxygen and lose consciousness. Dr. Morgan watched many of them sink to the bottom of the pool before divers pulled them to the surface. Once back on the deck, the unconscious sailors are rolled on their side. As soon as they're revived, an instructor shouts repeatedly: "Are you gonna quit?"

Sailors are given thirty seconds to answer or they're kicked out of the program. If they say they want to keep going, they're given another thirty seconds to recover and then are thrown back into the pool. It might sound sadistic, but the Navy is simply trying to identify those who are likely to survive the most dangerous missions. Through this grueling test, it finds sailors who refuse to give up, who can suppress the need to breathe, who trust that they'll be rescued if something goes wrong, and who are prepared to lose consciousness—or even die.

In another arduous test, Navy divers are taken three miles off the Gulf Coast at

Scuba students at the Naval Diving and Salvage Training Center undergo confidence training in a twelve-foot pool as part of their scuba certification course. Confidence training is one part of the students five-week course used to test their reaction to real-world emergency situations. ⌄

night and are given a target destination on the beach. Dumped in the water, the students submerge and are not allowed to surface until they reach their objective (similar to BUD/S training). To make the challenge even more stressful, a clock is running and the sailors aren't allowed to go deeper than twenty-five feet. Despite the tides and currents, they're also forbidden to swim parallel to the beach looking for their target. The penalty for breaking any of the rules is immediate expulsion from the course. Speed, efficiency, and accuracy are a critical part of their grade.

In this underwater navigation test, as in the pool exercise, Morgan again found that brain chemistry plays a huge role. The more NPY your body can pump into your system, the better you will perform.

Morgan also learned that the best underwater navigators release a lot of a natural steroid called DHEA (dehydroepiandrosterone), which buffers the effects of the stress hormone cortisol and helps the brain's hippocampus with spatial relationships and memory. Divers with the most NPY and DHEA graduated at the top of the class. Conversely, those with the lowest amounts performed poorly.

The Metronomic Heart

At POW camp and dive school, Morgan discovered a simple and accurate way of predicting who will survive and perform the best under extreme stress. You might call it the telltale heart. It starts with something called heart-rate variability, or HRV, the variations between beats. Healthy people have a lot of variability in the intervals between their beats, with their hearts speeding up and slowing down all the time.

It turns out that the best survivors have very little heart-rate variability. They have "metronomic heartbeats." In other words, their hearts thump steadily, like metronomes, with almost no variability between beats. The intervals between beats are evenly spaced. Morgan believes that a metronomic heartbeat is an easy way to detect good survivors and high neuropeptide Y releasers, which makes sense biologically because your brain stem, which controls your heartbeat, has a high density of neuropeptide Y.

Morgan analyzed the heartbeats of soldiers and sailors before they experienced major stress and found that the ones with metronomic heartbeats performed the best in survival school and underwater navigation

U.S. soldiers conduct a convoy through the snow-covered Salang Pass in the Parwan province, Afghanistan. The pass navigates through a winding road and tunnel through some of the harshest terrain in Afghanistan. Nearly two miles above sea level, the Salang Pass connects the Parwan and Baghlan provinces and serves as one of the routes to move cargo and supplies from the main logistics hub at Bagram Air Field, located in eastern Afghanistan, to Regional Command North. The Soldiers are with the 101st Sustainment Brigade. ⩔

Coronado, Calif—A Special Warfare Combatant-craft Crewman » (SWCC) candidate from Crewmen Qualification Training (CQT) Class 61 jumps into Coronado Bay for a floatation check. CQT is a 14-week course that teaches SWCC candidates the skills they need as members of the Naval Special Warfare Boat Teams, including navigation, craft maintenance and repair, towing, anchoring, and weapons. SWCC operate and maintain the navy's inventory of state-of-the-art, high-speed boats in support of special operations missions worldwide.

testing. They also performed the best in what's called close-quarters combat (CQC) training.

Morgan analyzed soldiers' heart rates right before they went into mock battle—while they were suited up in combat gear, waiting for a buzzer to ring that would send them running into a building to "neutralize" (kill) the enemy and rescue hostages. (They use "simunitions," simulated ammunition) The ones with metronomic heartbeats, Morgan concluded, neutralized more of the enemy and shot at fewer hostages and other non-threats.

But he also found that the metronomic effect is usually associated with early heart disease and even sudden death. Apparently the same body chemistry that allows people to survive under high stress does not translate into optimal heart health past the age of fifty. But without it, these elite forces might never have survived that long.

2

THE ELEMENTS OF SURVIVAL

"We all knew there was just one way to improve
our odds for survival: train, train, train. Sometimes, if your training is properly intense, it will
kill you. More often—much, much more
often—it will save your life."
—*Richard Marcinko, Navy SEAL*

The Survival Mind-Set

Why do some people with little or no survival training manage to prevail in life-threatening circumstances, while others with survival training die? The answer is mental fortitude, or the will to survive.

The key to survival is your mental attitude!

Psychology of Survival

Life-threatening situations create challenges that impact the mind. The thoughts and emotions that result can transform a confident, well-trained person into an indecisive, ineffective mass of neuroses and fears.

Obviously, none of us wants to fall apart when challenged. That's why it's imperative to familiarize yourself with the specific stresses associated with survival—and learn to understand and manage your own reactions to those stresses.

Stress

Whether we like it or not, stress is a condition that we're all familiar with. It's basically our reaction to the pressures of living. It can be defined as "a specific response of the body to a stimulus (such as pain) that interferes with the normal physiological equilibrium of an organism."

While stress is a condition most people seek to avoid, it also has benefits. For example, it provides a means of testing our strengths and values. It also helps us measure and develop our adaptability and flexibility. And, it can stimulate us to perform at higher levels.

Since we usually don't consider unimportant events to be stressful, stress can also be an excellent indicator of the significance we attach to a particular event. In other words, it tells us what we consider important.

All of us need some stress in our lives, but too much can be destructive. The goal, therefore, is to learn to manage stress so that it never becomes overwhelming. Too much stress leads to distress, which can adversely affect individuals and organizations.

Distress causes an uncomfortable condition that we try to escape and, preferably, avoid. Listed below are some common indicators of stress becoming distress:

Difficulty making decisions
Angry outbursts
Forgetfulness

Low energy levels

Constant worrying

Propensity for mistakes

Thoughts about death or suicide

Trouble getting along with others

Withdrawing from others

Hiding from responsibilities

Carelessness

Stress can be constructive or destructive. It can encourage or discourage, move us forward or stop us dead in our tracks. It can fill life with added meaning or render it meaningless. Stress can inspire us to perform at our maximum efficiency in a survival situation. It can also cause us to panic and forget all our training.

The key to your survival is how effectively you manage the inevitable stresses you will encounter. The survivor will work with his stresses instead of letting his stresses work on him.

Survival Stressors

Events—such as diving, shooting, skydiving, or being shot at—produce stress. The events themselves are not stress, but they produce it and are called "stressors." In other words, stressors are the cause, while stress is the response.

Once the body recognizes the presence of a stressor, it begins to prepare to protect itself—to either "fight or flee." First, the brain sends a message throughout the body. The body responds by releasing stored fuels (sugar and fats) to provide quick energy. Breathing rate increases to supply more oxygen to the blood. Muscle tension increases to prepare for action. Blood clotting mechanisms are activated to reduce bleeding from cuts. Senses become more acute (hearing becomes more sensitive, eyes widen, sense of smell sharpens) so that you are more aware of your surroundings. Heart rate and blood pressure rise to provide more blood to the muscles. This protective posture prepares you to cope with potential dangers. But it's impossible to maintain such a level of alertness indefinitely.

One stressor doesn't leave because another one has arrived. Stressors piggyback on one another. And the cumulative effect of minor stressors can add up to major distress, especially when they happen simultaneously.

As the body's resistance to stress wears down and the sources of stress continue or increase, the body becomes exhausted. At this point, the body loses its ability to resist stress or use it in a positive way, and signs of distress appear.

So anticipating stressors and developing strategies to cope with them are two important ingredients in the effective management of stress. Therefore, it's essential that a person in a survival setting be aware of the types of stressors he or she is likely to encounter.

Injury, Illness, or Death

In a survival setting, injury, illness, and death are potential dangers. It can be very stressful being alone in an unfamiliar environment and knowing that you could die from hostile action, an accident, or from eating something lethal.

Furthermore, illness and injury can add to stress by limiting your ability to maneuver, procure food and water, find shelter, and defend yourself. Illness and injury also add to stress through the pain and discomfort they generate.

It's only by learning to control the stress associated with one's vulnerability to injury, illness, and death that a survivor can muster the courage to take the risks associated with performing survival tasks.

Uncertainty and Lack of Control

Some people have trouble operating in settings where everything isn't preplanned and laid out. The only guarantee in a survival situation is that nothing is guaranteed. You'll be operating on limited information in a setting where you have little or no control over your surroundings. This uncertainty and lack of control can add to the stress of being ill, injured, or killed.

Environment

Even under ideal circumstances, nature is an unpredictable and potentially dangerous force. In a survival situation, you will have to contend with the stressors of possible capture by the enemy, extreme weather conditions, rugged terrain, and the variety of creatures inhabiting a particular area. Heat, cold, rain, winds, mountains, swamps, deserts, insects, dangerous reptiles, and other animals are just a few of the challenges that you might expect. Depending on how you handle the stresses of your environment, your surroundings can either provide a source of food, water, and protection, or can be a cause of extreme discomfort leading to injury, illness, or death.

Hunger and Thirst

Foraging can be another major source of stress. Without food and water, you will weaken and eventually die. Therefore, securing and preserving food and water will take on increased importance as your time in a survival setting increases.

Fatigue

Maintaining the will to survive is likely to get harder as you grow tired. In fact, it's possible to become so fatigued that the act of staying awake is stressful in itself. There are distinct advantages to facing adversity with others. The company of a teammate(s) can provide you with a greater sense of security and a feeling that someone is available to help if problems occur. In many survival situations, however, one is often called upon to rely solely on his or her own resources.

Summary

The survival stressors mentioned above are by no means the only ones you may face. Remember, what is stressful to one person may not bother another. Your experience, training, attitude, physical and mental conditioning, level of self-confidence, and survival mind-set will determine what you will find stressful in a survival environment. The object is not to avoid stress, but rather to manage the stressors of survival and make them work for you.

Now that you are armed with a better understanding of stress and the stressors common to survival, the next step is to examine how you are likely to react to the stressors you may face.

Natural Reactions

Human beings have been able to survive many major changes in their environment throughout the centuries. Their ability to adapt physically and mentally to a changing world has helped them thrive, while other species have gradually died off.

The same survival mechanisms that worked for our ancestors can help keep us alive today! Our challenge is to understand them better and anticipate how they will affect us in situations of acute danger.

Survival circumstances will produce psychological reactions in the average person. Let's examine some of the major reactions you might experience when faced with the survival stressors described above.

Fear

Fear is an emotional response to dangerous circumstances that has the potential to lead to death, injury, or illness. The anticipated harm need not be only physical. The threat to one's emotional and mental well-being can generate fear as well.

When trying to survive, fear can have a positive function if it encourages you to be especially cautious in situations where recklessness could result in injury. But fear can also immobilize you and cause you to become so frightened that you fail to perform activities essential to your survival.

It's perfectly normal to experience a certain degree of fear when placed in unfamiliar surroundings under adverse conditions. But it's imperative that you learn not to be overcome by these fears. Realistic training can help you acquire the knowledge and skills needed to increase your confidence and manage your fears.

Anxiety

Because it is natural to be afraid, it is also natural for us to experience anxiety. Anxiety can be an uneasy, apprehensive feeling we get when we anticipate or are faced with dangerous situations (physical, mental, and emotional). When used in a positive way, anxiety can motivate us to act to end, or at least master, the dangers that threaten our existence.

Look at it this way: If we were never anxious, there would be little motivation to make changes in our lives. The act of reducing your anxiety brings the source of that anxiety—i.e., your fears—under control. When it motivates us to take positive steps to relieve it, anxiety is helpful. But it can also have a devastating effect, overwhelming you to the point where you become easily confused and have difficulty thinking clearly. Once this happens, it becomes increasingly difficult to make good judgments and sound decisions. Therefore, it's important to learn techniques to calm anxieties and keep them in the range where they're helpful, not debilitating.

Anger and Frustration

Frustration arises when a person is continually thwarted in his or her attempts to achieve a goal. The goal of survival is to stay alive long enough until you are able to reach help or until help can reach you. To reach this goal, it's necessary to complete certain tasks with minimal resources. And it's inevitable that something will go wrong in trying to execute these tasks. Something will happen that's beyond your control. With one's life at stake, every mistake is magnified in terms of its importance.

Thus, sooner or later, you will have to cope with frustration when your plans run afoul. One byproduct of this frustration is anger.

As you would imagine, there are many events in a survival situation that can cause frustration or anger. Getting lost, damaged or forgotten equipment, the weather, inhospitable terrain, proximity to the enemy, and physical limitations are just a few.

Frustration and anger encourage impulsive reactions, irrational behavior, poorly thought-out decisions, and, in some instances, an "I quit" attitude (as people will sometimes avoid doing something they can't master). If harnessed and properly channeled, the emotional intensity associated with anger and frustration can help you productively answer the challenges of survival. Failure to do so can cause you to waste energy in activities that do little to further either your chances of survival or those of your teammates.

Depression

It takes a rare person not to feel some sadness when faced with the privations of survival. As this sadness deepens, we label the feeling "depression," which is closely linked with frustration and anger.

A person who is frustrated becomes angrier as he fails to reach his goals. If the anger fails to push you to succeed, frustration levels usually climb even higher, creating a destructive cycle between anger and frustration. As the cycle continues, one becomes increasingly worn down—physically, emotionally, and mentally. Finally, one starts to give up as his focus shifts from "What can I do?" to "There is nothing I can do."

Depression is an expression of hopelessness. There's nothing wrong with feeling sad as you think about your loved ones and remember what life is like back in "civilization." Such thoughts, as long as they remain temporary, can spur you to try harder and survive another day. On the other hand, if they become obsessive, you're likely to sink into a depressed state that saps your energy and, more important, your will to survive. It's imperative to resist falling into depression.

Loneliness and Boredom

Man is a social animal. We human beings generally enjoy the company of others. Very few people want to be alone *all the time!*

As you are aware, many survival situations involve isolation. This isn't necessarily a bad thing. Because loneliness and boredom can bring to the surface qualities you never knew you had, You might find that your imagination and creative abilities surprise you or that you have a hidden talent. Most important, you may tap into a reservoir of inner strength and fortitude you never knew you had.

Conversely, loneliness and boredom can lead to depression. In a survival situation you must find ways to keep your mind productively occupied and develop a degree of self-sufficiency. You must have faith in your ability to "go it alone."

Survivor's Guilt

The circumstances that have caused you to be in a survival setting are sometimes dramatic and tragic. You might be there because of an accident or a mission gone bad. Perhaps you were the only or one of a few survivors. While relieved to be alive, you might also be mourning the deaths of others who were less fortunate.

It's not uncommon for survivors to feel guilty about being spared while others die. This feeling, when used positively, has encouraged people to try harder to survive with the belief they were allowed to live for some greater purpose. Others have willed themselves to live so that they could carry on the work of those killed. Whatever reasoning you employ, it's critical to never undermine your will to survive. The alternative will only serve to compound the original tragedy. It's the survivor's imperative to survive.

Preparing Yourself

Your mission in any survival situation is to stay alive. As discussed, you should expect to experience a whole gamut of thoughts and emotions. These can either help you or lead to your downfall. Fear, anxiety, anger, frustration, guilt, depression, and loneliness are all possible reactions to the many stresses you're likely to encounter.

Now that you understand the psychological dynamics of survival situations, you need to prepare yourself to deal with them so you can further your ultimate interest, which is staying alive with honor and dignity. Know that the challenges of survival have produced countless examples of heroism, courage, and self-sacrifice.

Below are some tips to help you prepare yourself mentally:

Know Yourself

Strengthen your stronger qualities and work even harder to develop the areas where you know you're vulnerable.

Anticipate Fears

Begin thinking about what would frighten you the most if you were forced to survive alone. Train harder in those areas of concern. The goal is not to eliminate the fear, but to build confidence in your ability to function.

Be Realistic

Don't be afraid to make honest appraisals. See circumstances as they are, not as you want them to be. Keep your hopes and expectations within the parameters of your situation. Understand that unrealistic expectations only increase the probability of disappointment. Follow the adage, "Hope for the best; prepare for the worst." It is much easier to adjust to pleasant surprises than to cope with unexpected harsh realities.

Maintain a Positive Attitude

See the potential good in any situation. This not only boosts morale, it also stimulates your imagination and creativity.

Remind Yourself What Is at Stake

Failure to prepare yourself mentally for the stresses of survival can invite depression, carelessness, inattention, loss of confidence, poor decision-making, and giving up psychologically before the body does. Your life and the lives of your teammates may be at stake.

Train

Prepare yourself to cope with the rigors of survival. Whether you're engaged in military training or learning from life experience, understand that the skills you're acquiring will give you strength and confidence. The more realistic the training, the less overwhelming an actual survival setting will be.

Learn Stress Management Techniques

People under high levels of stress are more likely to panic if they're not well-trained physically and prepared psychologically to face life-threatening situations. While it's impossible to control specific survival circumstances, it is within our ability to monitor and manage our responses to those circumstances. Learning stress management techniques can significantly enhance your capability of remain calm and focused as you work to keep yourself and others alive.

Important techniques to develop include relaxation skills, time management skills, assertiveness skills, and cognitive restructuring skills (the ability to control how you view a situation).

S	**Size Up the Situation** **(Surroundings, Physical Condition, Equipment)**
U	**Use All Your Senses,** **Undue Haste Makes Waste**
R	**Remember Where You Are**
V	**Vanquish Fear and Panic**
I	**Improvise**
V	**Value Living**
A	**Act Like the Natives**
L	**Live by Your Wits, But for Now,** **Learn Basic Skills**

S -Size Up the Situation

If you are in a potential hostage situation, find a place where you can conceal yourself from the enemy. Security takes priority. Use your senses of hearing, smell, and sight to get a feel for your surroundings.

Size Up Your Surroundings

Determine the pattern of the area. Get a feel for what is going on around you. Every environment, whether forest, jungle, arctic, desert, water, or urban, has a rhythm or pattern. This rhythm or pattern includes animal and bird noises and movements and insect sounds. It may also include enemy traffic and civilian movements.

Size Up Your Physical Condition

Check any wounds you may have sustained and provide self-aid. Give first aid to your teammates if required. Take care to prevent further bodily harm. Stay hydrated. Dress appropriately for the environment.

Size Up Your Equipment

Check to see what equipment you have and what condition it's in.

Once you've sized up your situation, surroundings, physical condition, and equipment, you're ready to make your survival plan. In doing so, keep in mind your basic physical needs—water, shelter, and food.

U -Use All Your Senses; Undue Haste Makes Waste

During combat, any quick move or reaction made without thinking or planning can result in your capture or death. Don't move just for the sake of taking action. Consider all aspects of your situation before you make a decision and a move. If you act in haste, you may forget or lose some of your equipment or become disoriented so that you don't know which way to go. Plan your moves. Be ready to move out quickly without endangering yourself. Use all your senses to evaluate the situation. Note sounds and smells. Be sensitive to temperature changes. Be observant.

R -Remember Where You Are

Always know your position on the map and relate it to the surrounding terrain. If there are other people with you, make sure they also know their location. Pay close attention to where you are and to where you're going. Don't rely on others in the group to keep track of the route. Constantly orient yourself. During wartime, always try to determine how your location relates to—

The location of enemy units, hostile elements, and safe areas.

The location of friendly units.

The location of water sources.

Areas that will provide good cover and concealment.

V -Vanquish Fear and Panic

The greatest enemies in a combat survival and evasion situation are fear and panic. If uncontrolled, they can destroy your ability to make an intelligent decision. They may cause you to react to your feelings and imagination rather than to your situation. They can drain your energy and thereby cause other negative emotions.

I -Improvise

In the United States, we have all types of items available for our every need. Many of these items are cheap to replace when damaged. Our easy-come, easy-go, easy-to-replace culture makes it unnecessary for us to improvise much. This inexperience in improvisation can be an enemy in a survival situation.

Learn to use natural objects around you for different needs (i.e., using a tree limb as a lever). No matter how complete your survival kit may be, it will be insufficient and in time will run out or wear out. Your imagination and ingenuity must take over when that happens.

V -Value Living

Many of us have become creatures of comfort and dislike inconveniences and discomforts. What happens when we're faced with a survival situation with its stresses, inconveniences, and discomforts? This is when your will to live is vital. The experience and knowledge you've gained through life, training, and past experiences will have a bearing on your will to live. Stubbornness, or a refusal to give in to problems and obstacles, will give you the mental and physical strength to endure.

A -Act Like the Natives

The natives and animals of a region have adapted to their environment. To get a feel for the area, watch how the people go about their daily routine. When and what do they eat? When, where, and how do they get their food? When and where do they go for water? What time do they usually go to bed and get up? These actions are important to you when you're trying to avoid capture.

Animal life in the area can also give you clues on how to survive. Animals require food, water, and shelter. By watching them, you can find sources of water and food. Keep in mind that the reaction of animals can reveal your presence to the enemy. If in a friendly area, one way you can gain rapport with the natives is to show interest in their tools and how they get food and water. By studying the people, you learn to respect them, you often make valuable friends, and, most important, you learn how to adapt to their environment and increase your chances of survival.

L -Live by Your Wits and Learn Survival Skills

Without the right mind-set and proper training in survival, your chances of living through a survival and evasion situation are slight.

Learn these basic skills now. How you decide to equip yourself before deployment can impact whether or not you survive. You need to know about the environment to which you're going, and you must practice basic skills geared to that environment.

Practice basic survival skills during all training programs, exercises, and missions. Survival training reduces fear of the unknown and gives you self-confidence. It teaches you to *live by your wits.*

Pattern for Survival

Develop a survival pattern, one that includes water, food, shelter, fire, signals, and first aid. For example, in a cold environment, you'll need a source of water, traps or snares to get *food;* a *shelter* to protect you from the elements; *fire* for warmth and cooking; a means to *signal* friendly aircraft; and *first aid* to maintain health. *If injured, first aid has top priority* no matter what climate you're in.

Change your survival pattern to meet your immediate physical needs as the environment changes.

Project Planning

- Develop a safety and communications plan
- Anticipate problems
- Weigh production vs. safety
- Preplan for emergencies
- Organize available resources

Wilderness Emergency Management

- Stay in control of your situation
- Analyze any immediate threats to your safety
- Prevent dehydration and heat or cold injuries
- Eat
- Get off your feet when possible
- Remain positive
- Try to determine the length of your survival situation and prepare accordingly
- Protect and maintain your life
- Stay out of the elements when possible
- Seek or make a shelter
- Administer self-aid and first aid to teammates
- Protect your equipment
- Keep all equipment in your shelter
- Conserve your resources
- Do not throw anything away
- Inventory both man-made and natural materials available to you
- Use all available signaling equipment to give your status and location to friendly forces
- Rest when you can
- Stay alert
- Be patient

The Will to Survive—by Wade Chapple

One of the training venues the rescue center in Colombia offers to embassy folks is called "High Risk of Isolation." Essentially, the term "isolated" means" being separated from your unit or organization and friendly authorities." Anybody working outside of major population centers in Colombia can find themselves isolated and in need of assistance. In extreme circumstances, persons may find themselves isolated deep in the jungles of Colombia with their very survival totally dependent on their ability to help themselves.

A frequent question that comes up during this training venue is, "What is the most important key to survival when isolated in the Amazon jungle?" My response is always quick and to the point: the will to survive. Yes, knowing how to collect water and make it potable is important and yes, being able to make a fire, build shelter, and procure food are also important to survival. However, if one does not possess a fierce desire to survive, that person will likely perish if his/her rescue doesn't come quick.

This desire to survive is a form of mental toughness or psychology that enables a person in life-threatening survival situations to overcome the stresses that produce adverse emotional reaction and indecision.

Managing stress, controlling emotions, understanding one's reaction to particular challenges, and having the ability to suppress negative thoughts are all necessary traits that serve to make one mentally tough when facing life-threatening situations. The operator or individual who possesses these traits will also possess the will to survive.

Resilience: Can the Will to Survive Be Learned? —by John Bruce Jessen, Ph.D

Survival experiences span a multitude of settings. Climbers poised for a summit assault on a high alpine peak could be pinned down by severe weather, exhaust their supplies, and be faced with a survival experience. A tourist traveling through a vast desert might be accidentally left at a roadside stop, wander off to seek help, and become lost. The exhilaration of a pleasant afternoon sailing trip might tempt novice sailors into waters too demanding for their abilities. A military pilot flying a routine peace-keeping mission could be unexpectedly shot down, forced to make a sudden and shocking transition from a pilot secure in the cockpit to an evader in hostile territory. This same pilot could now find himself a captive facing exploitation and the threat of death. Some survival experiences are entered into almost voluntarily and some are thrust upon us. This spectrum of environmental variables is further complicated, if not largely determined, by the human element—the survivor.

Survivors vary with the same infinite randomness as do survival experiences. Environmental and human variables produce an incredible array of survival experiences. No matter how survival experiences vary, they all appear to share one critical element—the sense of individual loss of control. The degree of control loss will roughly determine the magnitude of the survivor's perceived distress. The conditions of one person's survival experience could very well produce merely an exciting adventure for another. An excellent example of this is the first unsupported crossing of the Antarctic

Continent by the Pentland South Pole Expedition. This expedition was led by Sir Ranulph Fiennes. For Sir Ranulph, navigating the high Antarctic plateau while pulling incredibly heavy sled loads in the extreme cold of catabolic winds was a seriously challenging and extremely rewarding exploration. However, if I were thrust into these same circumstances, it would be a survival experience of catastrophic proportions.

This comparison points to another constant in survival experiences. Individuals define survival experiences based on their own perceptual appraisal of the challenge they face. The ability to predict the eventual outcome of a survival situation is critical. It accounts for vast differences in human perception and performance. Prior to his expedition, Sir Ranulph read volumes of research on clothing, equipment, climatic conditions, and accounts of previous Antarctic explorers. He engaged in field trials which tested his equipment and tempered his physical and psychological strength. To that end, the majority of the peril he and his companion encountered, though very severe, was relatively predictable to him. Because I would never voluntarily participate in such a terrifying effort, I would not amass the qualitative life experiences and preparation necessary to cope effectively. I would feel frightened and inadequate. Each new and cruel circumstance would increase my feeling of vulnerability and helplessness. Sir Ranulph met each new challenge with measured optimism. His preparation and life experiences gave him realistic self-confidence. This unique will to persist and survive is known as resilience.

Sir Ranulph's genuine and hard-won confidence produced a healthy, optimistic approach to the challenges he faced. Though he experienced disappointment and fear, his optimism always won out. He was able to realistically predict a positive outcome for these challenges. For him, survival challenges served as cues which triggered his prepared mind and body to react in an adaptive and ultimately successful manner. The individual who is properly prepared enjoys a sense of control or composure. This realistic composure allows one to predict what will happen with a high degree of accuracy. The self-confidence which results yields an optimism that sustains the individual through disappointments and difficult times. This process produces resilience.

I have often heard survival instructors tell their students that survivors **must** have the "will to survive." They emphasize that survivors must

☆ **Norway - Navy Special Warfare reserve component sailors from SEAL team 18 maintain security during a direct action mission with the German Kommando Spezialkrafte (KSK) during cold response. cold response is a Norwegian exercise open to all NATO nations for winter warfare and joint coalition training.**

≈ **Virginia Beach, Va—A Navy SEAL platoon performs a land warfare exercise during a capabilities demonstration at Joint Expeditionary Base Little Creek, Va. The Naval Special Warfare community event was part of the 41st UDT/SEAL East Coast reunion celebration.**

"never give up;" that if they persist, they can survive. There are numerous stories of successful survival against the greatest of odds because survivors possessed the requisite resilience. However, there are also too many accounts of individuals relatively immobilized by a survival experience because they lacked adequate resilience. These individuals sat helplessly until rescued or released. In some unfortunate cases, they succumbed before rescue or release could be effected. In a significant number of these cases, it has been determined individuals possessed adequate means to endure and sustain life but lost "the will to survive" and perished. How do we account for these differences? Why do some individuals possess the necessary resilience to survive and others do not?

This contradiction provokes several critical questions, which I will address. The answers lead to a proven, scientific solution to this paradox, and a challenge to those engaged in the profession of survival and resistance training. I will use captivity as the survival experience on which to focus this examination.

What reactions do captives typically experience? Individuals who are thrust into captivity share universal reactions which include initial startle or panic, disbelief, denial of the reality of the situation, and a deep sense of vulnerability and helplessness. As the experience progresses from moments, to hours, and then days, captives experience anger and depression, asking themselves the pervasive question—"why me?" They often engage in bargaining, thinking to themselves, "If I get out of here, I will be a better person," or "I will dedicate my life to a noble cause," etc. As time wears on, their self-esteem erodes as the captive struggles with a devastating sense of loss of control. In prolonged captivity, usually more than six months, resignation and acceptance develop as anger and depression become less extreme. The captive attempts to structure life into surviving one day at a time. The magnitude of reactions captives experience is influenced by: (1) culture of the captor, (2) duration of captivity, (3) harshness of captivity, (4) support received from others if held with other captives, (5) innate predispositions, (6) commitment to the ideology or task which placed them in jeopardy initially, (7) maturity, (8) personal value systems, and (9) satisfaction with family relations during the precapture period. As these predictable reactions unfold, captives are urgently drawing upon their individual resources to meet this survival threat.

In what coping behaviors do most captives engage? Information from numerous captivity debriefs and firsthand accounts from survivors identify the most commonly used coping behaviors. Immediately upon captivity, individuals strive to counteract the shock and disbelief they are experiencing by reassuring themselves the experience will end soon. They disassociate themselves from their condition by sleeping or daydreaming. Most individuals engage in "magical thinking" related to rescues. They imagine some miraculous rescue occurring and "it all being over." They often fantasize their captors will realize it was all a mistake and release them. As hours turn into days, survivors work to establish some sense of control and predictability. They engage in communication efforts with other captives, establish physical fitness routines, and follow cleaning and personal hygiene rituals. They rely heavily on religious beliefs, prayer, and close family ties. When held together, captives develop a strong sense of group. Belonging to a pre-captivity group which is the object of their captors' exploitation often gives survivors a sense of being part of a "righteous" cause.

In further attempts to establish some control over their situation, captives engage in precaptivity life roles. Regardless of the degree of modification necessary, they think about or do things which remind them of their value as a person. They may repeatedly and neatly fold objects as perfectly as they can or offer first-aid or consoling companionship to other captives. They organize daily schedules, mentally construct "dream homes," and review past academic efforts as they strive for a sense of control and personal value. Reminiscing about past pleasant activities is typical. Most captives learn not to allow this to freely occur regarding family and loved ones because of the debilitating emotional effects. Rather, they restrict family reminiscences to a specific time and duration.

As survivors struggle to cope with captivity, they must constantly deal with waves of apprehension and fear. Apprehension and fear feed on a sense of uncontrollability. The less effective the captive perceives his coping efforts to be, the more the captive senses a loss of control. Captives strive to establish control by attempting to model remembered and present behavior of others who do not appear apprehensive or fearful. They create distractions from the dreadful reality of captivity. They try to dispute negative thoughts which persistently invade their thinking. The best defense in settings which produce apprehension and fear has proven to be repeated past experiences with anxiety and fear-producing settings similar to the present stressor, which were successfully overcome. Unfortunately, captivity is a stressor which doesn't come in gradual doses over time.

Although coping attempts are quite universal in their initial application, captives meet with varying degrees of success in applying them. Some individuals become overwhelmed almost immediately and show little or no optimism regarding their survival options. They seem unable to muster the necessary energy to continue coping. Others demonstrate initial energy in applying coping behaviors only to have this energy rapidly dissipate as survival demands increase. Others rebound quickly from their initial capture shock and persist in efforts to overcome new difficulties. These individuals demonstrate an ability to increase their coping skills and confidence over time to a highly resilient level. Research indicates almost all captives initially engage in similar coping behaviors; however, their resilience in continuing to do so as stress increases and novel threats arise varies greatly.

Given that individuals do not respond with equal effectiveness in captivity survival, what accounts for these varying degrees of resilience? A seemingly logical conclusion is that resilience is inborn, a result of genetic endowment. If this were true, those who were born with "the resilience gene" would be effective survivors and those without it would not. While there is evidence that some equally naive individuals do adapt to novel situations more quickly and effectively than others, sound scientific research shows that resilience is not the sole province of heredity. Fortunately, it can be acquired or learned. In order to understand how this learning takes place, we must first examine some important underlying dynamics. Studies have identified three elements which are necessary to produce resilience. None of the three alone is sufficient to produce the desired result, but when properly combined, they yield optimum resilience. The first element is talent. Talent is the basic raw material of intelligence and creativity which allows the survivor to conceive of and apply coping strategies. An average amount of talent is sufficient. The second element is desire or motivation. Once individuals recognize they are confronted with a survival situation, the desire to avoid or escape it is quite spontaneous. The third element is optimism. Survivors can have the desire to escape or avoid the situation and a basic idea of how to approach this, but if they lack optimism, their desire and talent alone will not result in the confidence and persistence necessary for resilient survival behavior. We know individuals possess varying levels of optimism and pessimism; still, the degree of optimism or pessimism a survivor demonstrates is significantly affected by learned behavior.

In order to understand the essential role optimism plays in resilience, it is necessary to contrast it with its opposite—pessimism. The prototypical pessimist believes bad events will last a long time. The pessimist believes bad events will undermine everything one does and are the result of some internal fault. Pessimists all have a normal endowment of talent and do not like aversive situations any more than optimists. However, hundreds of studies show pessimists give up easily when confronted with aversive situations which demand persistence to overcome. Prototypical optimists, on the other hand, believe bad events will be limited to a specific time and will affect only the specific situation in which they occur. Optimists also do not believe bad events are their fault. They view them as the result of external factors. In other words, optimists find temporary and specific causes for misfortune and do not blame themselves for their occurrence. This optimism, along with desire and talent, produces hope, confidence, and resilience.

≈ **Stennis Space Center, Miss—A Navy SEAL communicates with teammates during immediate action drills at the John C. Stennis Space Center. The drills are a part of the SEALs predeployment training. Navy SEALs are the maritime component of U.S. Special Operations Command and are trained to conduct a variety of operations from the sea, air, and land.**

The key variables in predicting whether a survivor will demonstrate optimism or pessimism are control and predictability. For the survivor, a sense of control yields predictability, which reduces stress, builds confidence, and allows the survivor to manage fear and anxiety. In scientific terms, we would say a sense of control yields less anticipatory arousal to potentially aversive stimuli. If the survivor cannot, to a reasonable degree, predict the outcome of the circumstances with which confronted, the survivor will become more and more pessimistic about a personal ability to cope. This pessimism will increase a sense of loss of control and undermine the survivor's ability to contend with the situation. As a result, a feeling of helplessness will develop. This helplessness, in turn, can become pervasive and effectively exclude present and even future coping behaviors from occurring.

This learned helplessness was demonstrated in a wide variety of animal studies. One such study was conducted on wild rats by Curt Richter at Johns Hopkins University. The rats were originally placed in a small metal cage. A sliding door was opened and the rat would see a dark opening (actually a black opaque bag). The rat would shoot through the opening, with retreat immediately cut off. The rat would then be gently pushed toward the end of the bag and simply held, through the walls of the bag, in the hand of the experimenter. Over 2,000 rats were held in this way and, despite their fierce and aggressive nature, none ever made an attempt to bite the experimenter through the bag. Many died simply from this restraint. Those that did not were placed in a large glass cylinder filled with water and forced to sink or swim. All died promptly on immersion in contrast to usual laboratory rats under the same circumstances who swam for incredible periods until eventually rescued. Electrocardiograph records were taken during this second phase and, contrary to the usual response of accelerated heart rate due to stress, which is consistent with a coping response, rats succumbing promptly had an immediate slowing of the heart rate. The implication that when fight or flight are not available, hopelessness and resignation follow with major psychophysiological responses that adversely affect survival. In this experiment, it was found that if wild rats were repeatedly held for just a few seconds and then released from this state, or if immersed in the cylinder of water very briefly and then "rescued," they showed no signs of giving up during later immersion. Instead, they continued swimming as long as domestic rats. A subsequent study with laboratory rats by John Wittrig and colleagues revealed that rats subjected to various kinds of stress in early life swam more than twice as long as a control group in which no stress was present during the same period of time.

As a result of these and many other studies, experimenters concluded that subjects who are permitted to survive an overwhelming situation develop an "immunity" to the "hopelessness" response. In numerous similar studies conducted on human subjects, the same effects occurred. Individuals who perceived they were helpless, because no amount of effort would effect a solution, ceased to cope. In contrast, those who were optimistic in their ability to eventually prevail persisted. This research evidence illustrates the difference between ineffective and effective resilient coping behaviors. When confronted by difficulties, those who have learned to be optimistic apply that optimism to their talent and desire and, consequently, persist in coping. Those who lack adequate experience or preparation, despite their talent and desire, are overcome, learn to be helpless, and cease coping. This learned helplessness effectively eliminates the confidence to persist in the face of adversity.

How does one acquire optimum survival resilience? There are two primary methods for acquiring resilience. First, it can be acquired through a process of naturally occurring education and graded exposures. I am confident it occurred in Sir Ranulph Fiennes' case. Allow me to extrapolate from Sir Ranulph's writings and describe how I conceive his learning occurred. His resilience was the result of an extremely thorough and comprehensive training process which occurred as a natural evolution of his life's pursuit of exploration. At some early stage in his life, he solidified the commitment to pursue exploration as his passion and profession. He read accounts of famous explorations. He talked to individuals who had participated in these harrowing adventures and gained insight from their experiences. He imagined what it would be like to experience these same kinds of adventure and accomplish similar conquests; he embarked upon initial, small-scale explorative forays of his own. These first efforts could have been in a local park or wilderness area close by. These initial efforts began to build a realistic, but not overwhelming, perspective of what exploration would be like, as well as the kind of training and equipment he would need. He honed his raw talent until it was well suited to his pursuit. His desire increased with each new successful conquest. He used this experience to begin a critical assessment of the magnitude of stresses and challenges he would be capable of confronting in the future. With this framework fixed in his mind, he embarked on only the most rigorous and demanding exploration challenges. He identified the limits of his endurance, his strong points, and areas where he needed improvement. He learned to distinguish conditions he could manipulate and those he could not change. He became extremely skillful in preparing for and conducting expeditions, confronting these stressful situations with optimism, regrouping when plans did not work out as well as he had hoped, and using those experiences to further refine skills. Eventually he was ready for "the ultimate test." This long sought-after exploration was fraught with the most severe of climatic conditions. He encountered extremely difficult and often unplanned physical and psychological obstacles. When these stressors occurred, he rebounded from their initial shock, gathered his confidence, and persisted with his quest to its eventual hard-won conclusion. Because of his desire, talent, careful preparation, confidence, and realistic optimism, he was eminently successful. His years of meticulous preparation provided him with incredible resilience. His optimum survival behavior led to his consummate performance on the Pentland South Pole Expedition.

If we return to our original vignette, where I, totally lacking Sir Ranulph's skill and excellent preparation, was to be thrust into similar Antarctic conditions, the results would be categorically different. I would be completely undone. I would have the desire to extricate myself. I have talent and some applied skills I could borrow from mountaineering; however, no doubt my optimism would fade fast and I would feel overwhelmed and helpless. Sir Ranulph met the challenge with a lifetime of preparation. For me, it would be an unplanned, unpredictable, and staggering survival experience.

Both methods for acquiring resilience have proven to be highly effective and systematic; both contain the same developmental process. If a survival

⌃ Special Warfare Combatant uses a red flare to signal an aircraft for a personnel recovery mission during a training exercise at Marine Corps Auxiliary Landing Field Bogue, N.C.

experience cannot be predicted, or if it involves a setting which does not lend itself to a natural development of situation-specific immunity, the second method is indicated. Where the first method is accomplished spontaneously over an extended period of time, the second can be implemented at any time, requiring relatively little time to complete, but still producing highly effective results. The second method is a specialized adaptation of stress inoculation training. Stress inoculation training is divided into three phases which approximate and greatly accelerate the natural development of the first method.

In the first or Conceptualization Phase, individuals develop a comprehensive understanding and reliable mental picture of the situation for which they are preparing. This is done by studying interviews and narrative accounts. Individuals are given a realistic, but not overwhelming, exposure to the situation which provides them an accurate assessment of the upcoming stressors. Then, based on this mental or cognitive picture, they conduct situational assessments and behavioral observations which build a sense of control and predictability. They learn which behaviors are productive and which will not be useful in coping with the identified stressor. They learn they are capable of enduring significant amounts of stress and how to handle the feeling of being overwhelmed when things do not go as planned.

Phase two is the Skills Acquisition and Rehearsal Phase. In this phase, individuals work on problem-focused solutions; they practice problem-solving activities by using coping behaviors in simulated and manageable realistic settings. Emotion-focused, palliative coping skills are also practiced, especially when the individual has to deal with unchangeable and uncontrollable external stressors like captivity. An extensive repertoire of coping responses are acquired and mastered gradually through multiple trials. When the mastery of effective coping skills has reached a proficient level, individuals engage in behavioral rehearsal, which is where effective self-regulated coping responses are demonstrated.

In phase three, the Application and Follow-Through Phase, individuals are exposed to more and more stressful situations where they apply their coping skills. Through these graded exposures, individuals reach a confidence level where, when placed in realistic "in-vivo" settings, they persist in optimum, resilient coping behavior.

This training has been empirically validated through numerous studies and is the training regimen of choice for preparing individuals to perform optimally in high-demand, stressful situations. It is easily adapted to a multitude of settings and produces the critical results we have been addressing. It is analogous to the traditional process of medical inoculation for the prevention of disease. This basic immunization inoculation rationale should be familiar to all of us. When I was young, polio was a severe physical threat to the health of our nation. We all had to receive an inoculation to hopefully prevent its devastating effects. We were injected with a carefully measured amount of polio vaccine. This amount of vaccine was calculated to cause our immune systems to react and recognize the antigen, but not become overwhelmed. The effect would be to simulate an attack from polio at a level which our immune system could react successfully and persist in the buildup of an ultimately protective capability. Once this was accomplished, should we become exposed to the full impact of the disease in the future, we would not be overwhelmed and helpless to it ravaging effects. Stress inoculation training produces analogous results for individuals confronting stressful life events. This training is remarkably adaptable to survival resistance training.

The ideal plan for incorporating stress inoculation training into resistance training is to utilize the Biderman resistance training model. This training model was originally designed in 1956 at the direction of US Department of Defense (DoD). At that time, the DoD tasked a working group of military experts, former POWs, and the most prominent behavioral scientists of the day to develop an ideal resistance training plan. Today, with minor, but significant scientific modifications, this model remains the state-of-the-art program for captivity and resistance training. This program was recently re-validated by four of the world's most eminent behavioral scientists (Joseph Mattarazzo, Ph.D., Charles Speilberger, Ph.D., Richard Lazarus, Ph.D., and Albert Bandera, Ph.D.). These highly respected authorities were unanimous in their evaluation of the Biderman model as being a superior learning process when compared to other more traditional DoD training models. The Bideman model has five components or interrelated phases of training:

1. Pre-Academic Captivity Exposure
2. Pre-Academic Captivity Exposure Debrief
3. Academics and Role-Play Laboratory
4. Post-Academic Detention
5. Post-Training Debrief

These training programs combine seamlessly to form an ideal resistance survival training program.

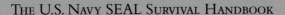

Stress Inoculation Training	Biderman Resistance Training Model
Conceptualization	**Pre-Academic Captivity Exposure**
Reliable Mental Picture	
Accurate Understanding of Conditions	
Initial Graded Exposure	
Assessment of Upcoming Stressors	**Pre-Academic Captivity Exposure Debrief**
Skills Acquisition and Rehearsal	**Academics and Role-Play Lab**
Work on Problem-Focused Solutions	
Practice Problem-Solving Activities	
Work on Emotion-Focused Coping Skills	
Develop Extensive Repertoire of Coping Responses	
Engage in Graded Exposures	
Application and Follow-Through	**Post-Academic Detention**
Exposure to Most Realistic In-Vivo Exposure	
Reinforce Appropriate Coping Behaviors	**Post-Training Debrief**

Who should receive this special application of stress inoculation training? Of all possible groups, this training is particularly applicable and easily adaptable to military organizations involved in survival resistance training. Over the past six years, this training method has proven to be extremely effective in real-world captivity incidents for the United States, Department of Defense, Joint Special Operations Command high-risk operators. Most recently, Dr. Jim Mitchell, Survival, Evasion, Resistance, and Escape (SERE) Psychologist at the US Air Force Survival School, compared a specialized application of stress inoculation resistance training with a more traditional resistance training approach on aircrew members. The results of his research show consistently higher levels of confidence in perceived ability to adhere to the Code of Conduct (in captivity) for aircrews receiving stress inoculation training as compared to aircrews who were trained by the more traditional method. Captivity occurs with great regularity for military personnel as compared with individuals in the civilian population. This is particularly the case for certain high-risk units. Because captivity is a fortuitous event, it is unrealistic to assume these high-risk operators will be exposed to the first naturalistic method of acquiring resilience.

Captivity is not a naturally occurring life event. Unlike exploration and high adventure wilderness sports, which tend to prepare participants for their unique potential survival difficulties, captivity is not a sought-after endeavor. It would be an extremely unfortunate person whose life had prepared him thoroughly for captivity. Nonetheless, military men and women consistently become captives. We have examined the consequences of thrusting unprepared individuals into extreme survival settings. We know that ideal levels of resilience in survival experiences are not inborn; they are situation-specific; if not acquired in a naturally-occurring manner, resilience must be learned. If resilience is not learned prior to captivity, resistance behavior will be much less effective or possibly inadequate. The consequences of this lack of resilience can be devastating to the captive, the captive's government, and comrades. Without resistance training, captives must endure one of the most extreme survival circumstances possible with little or no relevant knowledge, skills, or realistic confidence. With proper training, captivity survival can be proactive. Captives will be more capable of protecting information and themselves. They will survive in a manner which preserves their ability to fight again effectively instead of rendering them less prepared for future conflict as a result of learned helplessness.

The military services of many nations recognize the need to prepare their high-risk operators for the possibility of captivity and provide training for them. I have participated in several of these training programs and, without question, they are staffed with talented instructors who provide excellent training. All of these training programs strive to give the best possible preparation to their students. Those who are placed in harm's way in the service of their nations must be prepared. With minimal modification and staff training, these excellent programs can be even more effective. Those of us involved in the profession of survival training should constantly seek ways to improve students' preparation. Incorporating these empirically proven principals in a resistance-specific application of stress inoculation training will significantly improve a captive's ability to survive. The result will be optimum survival resilience.

3

JUNGLE SURVIVAL

"We shall draw from the heart of suffering itself

the means of inspiration and survival."

—*Winston Churchill*

⩘ **Tiger**

⩘ **Boa constrictor**

⩘ **Jungle snake**

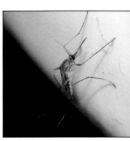

⩘ **Mosquito**

⩘ **Jaguar**

⩘ **Poison dart frogs**

Jungles, or rain forests, are lush, green areas teeming with life of all shapes and sizes. Although they only cover about two percent of the earth's surface, they're home to fifty percent of all plants and animals. If that doesn't describe how vital they are, consider this: A four-square-mile (ten-square-kilometer) area of a rainforest can contain as many as 1,500 plants, 750 species of trees, 400 species of birds, and 150 species of butterflies.

The good news about the jungle is that water and food are plentiful. The bad news is twofold:

1. The jungle's thick overhead canopy makes it nearly impossible for anyone to spot you.
2. There are lots of things that can kill you, including mosquitoes carrying malaria, small brightly colored poison dart frogs, snakes, poisonous plants, and even jaguars and tigers.

Stay or move away from snakes, particularly boa constrictors, coral snakes, and fer-de-lances. Try to avoid insects. They can cause serious allergic reactions. Learn from the natives, who rub garlic on themselves to ward off bugs and snakes.

If you encounter a jaguar, never run from it. Walk toward it while shouting and clapping.

Types of Jungles

Tropical areas can be described as:

- Secondary jungles
- Tropical rain forests
- Semi-evergreen seasonal and monsoon forests
- Tropical scrub and thorn forests
- Tropical savannas

- Saltwater swamps
- Freshwater swamps

Secondary Jungles

Similar to rain forests. The difference is that sunlight does penetrate to the jungle floor in secondary jungles. Such growth is typical along riverbanks, on jungle fringes, and where people have cleared the rain forest.

Tropical Rain Forests

Found across the equator in the Amazon and Congo basins, parts of Indonesia, and several Pacific islands. Rain forests are characterized by their climate, which includes about three and one half meters of rainfall throughout the year, and temperatures ranging from about 90 degrees Fahrenheit during the day to 70 degrees Fahrenheit at night.

Rain forests typically have five layers of vegetation:

1. Jungle trees rise from buttress roots to heights of sixty meters.
2. Smaller trees produce a canopy so thick that little light reaches the jungle floor.
3. Seedlings struggle beneath them to reach light and masses of vines and lianas twist up to the sun.
4. Ferns, mosses, and herbaceous plants push through a thick carpet of leaves.
5. A large variety of fungi grow on leaves and fallen tree trunks.

Since little light reaches the jungle floor, there is little undergrowth to hamper movement. But dense growth limits visibility to about fifty meters, making it easy to lose your sense of direction. And the thick canopy makes it extremely difficult to be spotted by aircraft.

Semi-Evergreen Seasonal and Monsoon Forests

Found in parts of Columbia, Venezuela, and the Amazon basin in South America; in portions of coastal southeast Kenya, Tanzania, and Mozambique in Africa; and in northeastern India, much of Burma (Republic of the Union of Myanmar), Thailand, Indochina, Java, and parts of other Indonesian islands.

These forests are characterized by two stories of tree strata. Those in the upper story average eighteen to twenty-four meters while those in the lower story average seven to thirteen meters. The diameter of both strata of trees averages one-half meter, and their leaves fall during seasonal droughts.

≫ **Jungle trees**

With the exception of sago, nipa, and coconut palms, the same edible plants grow in these areas as in the tropical rain forests.

Tropical Scrub and Thorn Forests

Found on the west coast of Mexico, Yucatan peninsula, Venezuela, and Brazil; on the northwest coast and central parts of Africa; and in Asia, Turkestan and India. They have five main characteristics:

1. There is a clearly defined dry season.
2. Trees are leafless during the dry season.
3. Fires occur frequently during the dry season.
4. The ground is bare except for a few tufted plants in bunches; grasses are uncommon.
4. Plants with thorns predominate.

When traveling through these areas, keep in mind that while plants are abundant during the rainy season, you will find it hard to obtain food plants during the dry season.

Tropical Savannas

Found within the tropical zones in South America and Africa, including parts of Venezuela, Brazil, and the Guianas in South America; and the southern Sahara (north-central Cameroon, Gabon, and southern Sudan), Benin, Togo, most of Nigeria, northeastern Zaire, northern Uganda, western Kenya, part of Malawi, part of Tanzania, southern Zimbabwe, Mozambique, and western Madagascar in Africa. They looks like broad, grassy meadows with trees spaced at wide intervals.

Saltwater Swamps

Found in West Africa, Madagascar, Malaysia, the Pacific islands, Central and South America, and at the mouth of the Ganges River in India. Common in coastal areas subject to tidal flooding, saltwater swamps have tides that can vary as much as twelve meters and have a large variety of hostile creatures, from leeches and insects to crocodiles and caimans. Avoid these swamps if possible.

Mangrove trees thrive in these swamps, and can reach heights of twelve meters. Their tangled roots are an obstacle to movement. Visibility in this type of swamp is poor.

Freshwater Swamps

Found in low-lying inland areas and characterized by masses of thorny undergrowth, reeds, grasses, and occasional short palms that reduce visibility and make travel difficult. There are often islands that dot these swamps, allowing you to get out of the water. Wildlife is abundant.

Familiarize Yourself with Your Environment

Knowledge of field skills, the ability to improvise, and the application of the principles of survival will increase your prospects of survival. Don't be afraid of being alone in the jungle, as fear can lead to panic. And panic can lead to exhaustion and decrease your odds of survival.

One of the worst aspects of a jungle is the density of the vegetation, which makes it difficult to navigate. To get a better viewpoint of your surroundings, look for a high area that is not obstructed or climb a tree and get above the canopy. If you see depressions in the jungle where one side is higher than the other, that could be a river, which means civilization may be nearby.

Nature will provide water, food, and plenty of materials for building shelters.

Everything in the jungle thrives, including disease germs and parasites that breed at an alarming rate.

Indigenous peoples have lived for hundreds of years by hunting and gathering. However, it will take a non-native significant time to get used to the conditions and activity of tropical survival.

Weather

Weather in a jungle environment can be harsh. High temperatures, heavy rainfall, and oppressive humidity characterize equatorial and subtropical regions, except at high altitudes. At low altitudes, temperature variation is seldom less than 50 degrees Fahrenheit and is often more than 95 degrees Fahrenheit. At altitudes over 1,500 meters, ice often forms at night. The rain has a cooling effect but when it stops, the temperature soars.

Rainfall can be heavy, depending on the season, often accompanied by thunder and lightning. Sudden rain beats on the tree canopy, turning trickles into raging torrents and causing rivers to rise. Jut as suddenly, the rain stops. Violent storms may occur, usually toward the end of the summer months.

Hurricanes, cyclones, and typhoons develop over the sea and rush inland, causing tidal waves and devastation ashore. In choosing bivy sites, make sure you're located above potential flooding. Prevailing winds vary between winter and summer. During the dry season rains falls only once a day, while the monsoon season has continuous rain.

Tropical days and nights are of equal length. Darkness falls quickly and daybreak is just as sudden.

Immediate Considerations

Because of the thick canopy found in most jungle areas, it's unlikely you'll be spotted from the air and rescued. You'll probably have to travel to reach safety.

If you're the victim of an aircraft crash, the most important items to take with you from the crash site are a machete, compass, first-aid kit, and a parachute or other material for use as mosquito netting and shelter.

1.   Take shelter from tropical rain, sun, and insects. Malaria-carrying mosquitoes and other insects are immediate dangers.
2.   Do not leave the crash area immediately, as rescuers may be looking for you. If you do decide to leave, don't do so without carefully marking your route. Use your compass and know what direction you are taking.
3.   In the tropics, even the smallest scratch can quickly become dangerously infected. Promptly treat any wound, no matter how minor.

Water

The good news is that if you're trapped in a tropical environment, the chances are that water is abundant—in the form of springs, rivers, and lakes. But initially you may have trouble finding it. Often you can get nearly clear water from muddy streams or lakes by digging a hole in sandy soil about one meter from the bank. Water will seep into the hole. Make sure to purify any water obtained in this manner.

Animals as Signs of Water

Grazing animals such as deer are rarely far from water and usually drink at dawn and dusk. Look for converging game trails, which will often lead to water.

Grain-eating birds, such as finches and pigeons, are never far from water. They also drink at dawn and dusk. When they fly straight and low, they are heading for water. When returning from water, they are full and will fly from tree to tree, resting frequently. Do not rely on water birds, or hawks,

A water-purifying table at a FARC (Fuerzas Armadas Revolucionarios de Columbia) camp. Members of this notorious terrorist group throw river water on top of a sand-base and charcoal-topped bed that funnels into large plastic containers held below. This is how they purify water! This pic was taken at a hostage-holding camp somewhere in the jungles of Colombia ⩢

eagles, and other birds of prey. Water birds fly long distances without stopping, and birds of prey get liquids from their victims.

Insects, especially bees, are often good indicators of water. Bees seldom range more than six kilometers from their nests or hives and usually will have a water source in this range. Ants also need water. A column of ants marching up a tree is going to a small reservoir of trapped water. Most flies stay within one hundred meters of water, especially the European mason fly, easily recognized by its iridescent-green body. Human tracks will usually lead to a well, borehole, or soak. Scrub or rocks may cover it to reduce evaporation. Replace the cover after use.

Water from Plants

1. **Vines** with rough bark and shoots about five centimeters thick can be a useful source of water. The poisonous ones yield a sticky, milky sap when cut. Nonpoisonous vines will give a clear fluid. Some vines cause a skin irritation on contact; therefore, let the liquid drip into your mouth, rather than put your mouth to the vine. Preferably, use some type of container.
2. **Roots**. The water tree, desert oak, and bloodwood have roots that grow near the surface. Pry them out of the ground and cut them into thirty-centimeter lengths. Remove the bark and suck out the moisture, or shave the root to a pulp and squeeze it over your mouth.
3. **Palm Trees**. Buri, coconut, and nipa palms all contain a sugary fluid that is very good to drink. To obtain the liquid, bend a flowering stalk downward and cut off its tip. If you cut a thin slice off the stalk every twelve hours, the flow will renew, making it possible to collect up to a liter per day. Nipa palm shoots grow from the base, so you can work at ground level. On grown trees of other species, you may have to climb them to reach a flowering stalk. Coconut milk has a large water content but may contain a strong laxative in ripe nuts. Drinking too much of this milk may cause you to lose more fluid than you drink.

Water from Condensation

Sometimes, rather than digging for roots, it's easier to let a plant produce water for you in the form of condensation. If you tie a clear plastic bag around a green, leafy branch, water in the leaves will evaporate and condense in the bag. You can get the same result from placing cut vegetation in a plastic bag.

≈ **A FARC kitchen in the jungle**

≈ **Banana tree**

≈ **Breadfruit**

Food

In the jungle, it may not be worth wasting energy hunting animals or setting traps since there are typically plenty of plants to eat. Take the time to familiarize yourself with edible plants of the region in which you are traveling before you set out on a trip. But if you're in the jungle and not sure a particular plant is edible or not, avoid it! Many plants in the jungle are poisonous.

Tropical Zone Food Plants

Unless you can positively identify edible plants, it's safer to begin with palms, bamboos, and common fruits. The list below identifies some of the most common foods:

- Bael fruit (*Aegle marmelos*)
- Bamboo (various species)
- Banana or plantain (*Musa* species)
- Bignay (*Anti esma bunius*)
- Breadfruit (*Artrocarpus incisa*)
- Coconut palm (*Cocos nucifera*)
- Fishtail palm (*Caryota urens*)
- Horseradish tree (*Moringa pterygosperma*)
- Lotus (*Nelumbo* species)
- Mango (*Mangifera indica*)
- Manioc (*Manihot utillissima*)
- Nipa palm (*Nipa fruticans*)
- Papaya (*Carica papaya*)
- Persimmon (*Diospyros virginiana*)
- Rattan palm (*Calamus* species)
- Sago palm (*Metroxylon sagu*)
- Sterculia (*Sterculia foetida*)
- Sugarcane (*Saccharum officinarum*)
- Sugar palm (*Arenga pinnata*)

≈ **Lotus**

≈ **Mangos**

≈ **Papaya tree**

≈ **Persimmons**

≫ **Sugarcane**

≫ **Palm tree**   ≫ **Taro**   ≫ **Figs**   ≫ **Rice**

- Sweetsop (*Annona squamosa*)
- Taro (*Colocasia* and *Alocasia* species)
- Water lily (*Nymphaea odorata*)
- Wild fig (*Ficus* species)
- Wild rice (*Zizania aquatica*)
- Wild yam (*Dioscorea* species)

Rain forests are also full of biologically active compounds, many of which you can use for food or medicine. You can eat edible tubers such as potato, yucca, and boniato, but be sure you can distinguish edible tubers from poisonous ones.

≫ **Palm frond hut**

Shelter

Local weather and predators will determine your shelter needs. Do you need warmth or just a roof? Are there animals or insects that you need avoid? Falling trees and branches are the biggest killers of people in the jungle, so if you have to make a bivy or a camp, find clear ground.

The Maya use cohune palm fronds to build thatched roofs on their huts. These roofs will withstand rain and wind and can last up to fifteen years.

Navigation

Trekking in the jungle can be hazardous. Watch where you tread. Step over a log or grab a vine without looking, and you could get stung or bitten.

≫ **A meeting place set up in the Amazon jungle**

⌃ **Jungle canopy view**

With practice, movement through thick undergrowth and jungle can be done efficiently. Always wear long sleeves and long pants to avoid sunburn, rashes, cuts, and scratches.

To move easily, you must develop "jungle eye," which means that you shouldn't concentrate on the pattern of bushes and trees immediately in front of you. Instead, focus on the jungle further out and find natural breaks in the foliage. Try looking *through* the jungle, not at it. Stop occasionally to examine the jungle floor. Look for game trails that you can follow.

Stay alert and move slowly and steadily through dense forest or jungle. Stop periodically to listen and gain your bearings. If available, use a machete to cut through dense vegetation, but don't cut unnecessarily or you'll quickly wear yourself out. If using a machete, stroke upward when cutting vines to reduce noise, since sound carries long distances in the jungle. Use a stick to part the vegetation. A stick will also help dislodge biting ants, spiders, or snakes. Do not grasp at brush or vines when climbing slopes. They may contain irritating spines, black palm, or sharp thorns.

Many jungle and forest animals follow game trails. These trails wind and cross but frequently lead to water or clearings. Use them if they lead in the general direction that you want to go.

In many countries, electric and telephone lines run for miles through sparsely inhabited areas. Usually, the right–of–way is clear enough to facilitate relatively easy travel. When traveling along these lines, be careful as you approach transformer and relay stations.

The sun and stars are always reliable navigational tools

Protect Your Clothing—by Wade Chapple

If you find yourself in a life or death struggle for survival and you do not have your second or third line gear with you, it is likely that you're wearing your only available clothing. In the jungle, understand that your clothing will wear out quickly, so take time to preserve the clothing available to you. Once it's gone, you'll become more exposed to the elements and thus more susceptible to injuries and disease.

Main Items:

T-shirts/undershirts: If you have one of these, I recommend you take it off and stow it away during daytime movement. You can remove your outer shirt and put on the t-shirt/undershirt once you've finished moving for the day. Changing out of sweaty outerwear and into a t-shirt or undershirt will be refreshing while affording you an opportunity to wash and dry your outer shirt.

Underwear: Don't wear underwear in the jungle, as they cause chaffing and rashes. If you do have underwear, wash it at the first opportunity and stow it away except to use as a washrag or bandage for blistered feet, lacerations, infected insect bites/stings, and so on. A clean pair thrown over your head at night might just prevent the mosquitoes from stinging your face, ears, and neck.

Outer clothing: If you find yourself alone in the jungle, hopefully you're already dressed appropriately for facing possible survival situations. Your pants and shirt should be loose fitting and have multiple cargo pockets where you can stow needed items (first line gear) such as a compass, GPS, signal mirror, and so on. Outer clothing will protect you from the vines, thorns, rocks, insects, sun, and other things you might brush against or face while moving through the jungle. Conversely, it is your outer clothing that is most subject to damage. Therefore, you must take the time to inspect and care for your outer clothing, ideally once a day. If your clothing rips, take the time to repair it as soon as possible; if you wait too long, that tear might become a significant rip that will ultimately turn your outerwear into a non-wearable rag!

Try to wash your outerwear at least once every two days. Simply immerse your outerwear in a water source and gently wring out the water. Repeat this a few times to rid the outerwear of caked-on dirt, sweat, and body salt. Dry your outerwear in the shade by laying the items on the ground or by hanging them on limbs or vines. Never dry clothing in direct sunlight, as doing this will weaken the fabric.

Boots: Leather footwear rots quickly in the jungle. Modern military-issue boots tend to hold up better than the old-style jungle boots, but the fabric on them still begins to weaken in a jungle environment. In order to take care of your boots, immerse them in water at least once every two days, preferably in the late afternoon when you have stopped moving for the day. To prevent inadvertent punctures or rips, rub your boots only with your hands to remove caked-on mud. While washing your boots, keep the laces laced and the insoles inside of the boots. After gently washing your boots, inspect them for damage, and then remove the laces and insoles to dry these separately. It's best to place two sticks into the ground and hang your boots upside down, one on each stick (the stick goes inside of the boot). Hang your bootlaces above the ground to air-dry and wring out your insoles before hanging these to air out and dry. Again, don't dry your boots and accessories in direct sunlight! Before moving the next day, reassemble each boot and rub moist mud all around the outside fabric in order to give the boots a protective covering. Finally, if you have animal fat in your possession, you can rub some of this fat into the fabric of your boots as well in order to prevent them from drying out and cracking.

Socks: If your situation allows, wash and dry them every day when you've finished walking. Your socks need to be washed and aired out daily to reduce the effects of flesh-eating fungus. In time, your socks won't have much fabric left on them, but don't worry. They can be used as sweat bands while walking. While your socks are drying, try to air out/dry out your feet as well.

Hats: Wash out dirt, grease, and salt daily by immersing your hat in water and gently wringing it out. Repeat the process as necessary. Hang in the shade to dry. Your hat will protect you from prolonged exposure to sunlight and consequential burns. If you wear a brimmed jungle or "floppy" hat, pin about a half-dozen safety pins on the brim for emergency use. You can also affix fishing hooks for use in emergency situations.

Remember, the clothes you have on you may be all you have. When they're gone or unusable, your suffering will increase. So take care of what you have in a survival situation.

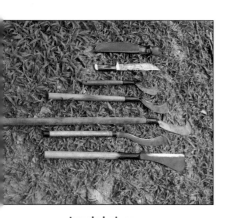

⋏ **Jungle knives**

Tropical Jungle Survival Kits—by Dave Williams

When traveling in, around, or over jungle areas, ensure that your survival kit contains the following essential items:

- Water purification tablets (iodine)
- Plastic sheet (for building water stills)
- Non-lubricated condoms (water storage)
- Fire starters
- Butane or similar lighter
- Metal match (magnesium fire stick)
- Waterproof matches (in waterproof container)
- Kindling (Vaseline-coated cotton balls, pre-charred cotton material, natural materials)
- Small magnifying lens
- Knife (sturdy fixed-blade knife with a four- to six-inch blade)
- A smaller knife for more delicate work (folding multitool, but be selective. Most tools are useless for survival situations.)

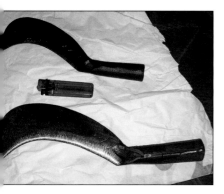

⋏ **Thai jungle knives: They're made of high-carbon steel that really holds an edge. They have a bit of weight to it, too, so when you put this on the end of a stick and swing it, you can really cut through stuff.**

- Snare wire
- Signaling mirror
- Compass
- Fish hooks and line
- Tetracycline tablets for diarrhea or infection
- Antibacterial ointment packets
- Solar blanket
- Surgical blades
- Butterfly sutures
- Insect repellent
- Needles and thread
- A mobile phone + extra battery, though larger and certainly more expensive, satellite phones have better coverage
- An EpiPen if you're allergic to insect stings
- Life-sustaining personal medication

Tropical Coastal (but not lost at sea) Survival Kit—Add-On Essentials

- Solar still material
- Additional fishing gear
- Heavier fishing line (Braided line is very strong, but visible. Bring monofilament also.)
- Lead or other metal weights
- Lures (jig and spoon-type lures pack well)
- Large bandana or piece of cloth for sun protection
- Collapsible sunglasses
- Lip balm with high SPF rating

Optional Coastal Kit Items—Space-permitting

- Hand-operated emergency desalination water pump
- Strobe light
- Signaling flares

Optional Land and Coastal Kit Items—Space-permitting

- Small water-purification pump
- Candle ("trick" birthday candles work well)
- GPS with mapping
- Whistle
- Tube tent or light hammock

- Lightweight poncho
- Aluminum foil
- Small flashlight and extra batteries
- Food bars/energy bars
- Parachute cord (550 cord)
- Sun protection
- Cooking kit
- Additional first-aid material
- Frog/fishing spear head
- Surgical tubing and slingshot pouch
- Assorted cable ties (many uses, from repairing broken gear to make-shift lashing)

What You Don't Need

- Rambo-style knife
- Multitools with everything imaginable
- Hatchet
- Knife-sharpening stone (nature provides plenty of natural stones that work just fine)
- A shovel
- Toilet paper

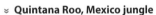
Quintana Roo, Mexico jungle

Anytime you venture off into the jungle or rain forest, you should carry the essentials: fire-making tools, a knife, a few water purification tablets and since you're probably going to be around water, some fishing line, sinkers, hooks, and maybe a lure and a map and compass.

How to Carry a Jungle Survival Kit

Waist packs work well. You carry your knife on the belt, too. Anything that can be destroyed by getting wet should be carried in a waterproof container. Look for one of several brands of waist packs that are waterproof already. Naturally, fire-making tools (lighters, matches, kindling, etc.) should be kept completely dry. Double-packing fire material is always a good idea. Even in wet conditions, if you have dry kindling and can locate somewhat dry tender, you can start a fire.

If you're on the sea or in a river, you can often carry most of your survival kit in your PFD (personal flotation device, i.e., life jacket) pockets. The remainder can be in a waist pack.

Teddy Roosevelt's Amazon Adventure

Theodore Roosevelt has the distinction of being the only president who became an Amazon explorer.

In 1913 Teddy and his son, Kermit, set off to Brazil to explore a recently discovered river deep in the Amazon jungle. Their adventure became known as the Roosevelt-Rondon Scientific Expedition.

The expedition encountered problems from the start. Disease-bearing insects attacked members of the group, leaving them in a constant state of sickness, with festering wounds and high fevers. The heavy dug-out canoes provided were unsuitable to the constant rapids and were often lost, requiring days to build new ones. The food provisions were ill-conceived, forcing the team on starvation diets. Native Indian cannibals (the Cinta Large) shadowed the expedition and were a constant concern. Also, one of the local guides murdered another; a third was killed in the rapids.

⌃ **A satellite image of part of the Amazon river.**

Teddy almost died from a wounded leg that became infected. His death five years later was blamed in part on health problems he developed during the expedition. But he survived and accomplished his goal of reaching the newly discovered river, which was later named after him.

Knives with hollow handles are popular for carrying survival gear, but they present two problems. First, the hollow handle offers a limited amount of space. Second, hollow handles are weaker. And you're going to want a survival knife that is extremely sturdy. That doesn't mean that it has to be heavy or a massive Rambo style knife, but it should be well made.

If you have the means to cut, pry, or split wood, coconuts, or whatever without using your knife, do it. Save your knife for those times when you truly need it.

To use your survival knife as a hatchet, hammer the back of the blade with a club-like piece of wood. This is much more efficient than hacking at the wood. When splitting bamboo, this method works wonderfully. In fact, there's no better way to do it.

Crossing Jungle Rivers—by Wade Chapple

One of the most difficult problems you'll encounter when you're moving under reduced visibility is what to do when you encounter rivers.

Piranhas, electric eels, crocodiles, and various other harmful creatures not-withstanding, the winding nature of jungle rivers can cause emotional stress, frustration and exhaustion to the already stressed person who is simply trying to survive and make it back to civilization.

≈ **Jungle terrain**

≈ **Quintana Roo, Mexico river**

Imagine slogging your way through the jungle and finding a roaring river in your path. You summon your strength and courage, and you cross it. Now you're on the other side, elated and exhausted. You move on. Five hundred meters later you run into what appears to be the same roaring river again!

Now repeat this process several times in the course of a day, then multiply it with several days of movement.

Having to cross a particular jungle river once is difficult enough. But having to cross the same river multiple times can prove to be more than a frustrating. However, if you understand the nature of jungle rivers and accept the fact that you are going to remain wet during most of your survival experience, there are ways to minimize the risks and emotional frustrations of jungle survival and onward movement.

There are three basic characteristics of jungle rivers:

1. Jungle rivers never run in a straight line. Instead, they generally wind back and forth like a snake.
2. What seems to be a principal river may be nothing more than an estuary or offshoot of the water's main body. Getting caught in the middle of these can be a frustrating and even life-threatening endeavor.
3. Although it may seem that a jungle river frequently reverses its course, it is actually moving in one general direction, which is downhill and toward the sea.

Follow these three basic rules:

Determine which direction you need to travel to reach civilization. Your choices are: with the current (downhill) or against it (uphill). Remember, rivers originate in the mountains and flow downhill toward the sea.

1. Estimate the general direction that the river is flowing. You can do this by walking parallel to the river and taking frequent azimuth recordings (as you would when navigating at every bend in the river) to learn the river's general direction. In fact, you should repeat this process often during your trek. Make sure you're already traveling in the direction you've chosen to take (upriver or downriver).

2. Once you have figured out the river's general direction, move away from one of the river's bends and turn to continue walking in the general direction you have previously determined—upriver or down. If the river is situated to your left, always keep the river to that side of you. If the river is to your right, keep it to your right. If you continue to run into the river, move farther away from it, then proceed in the general direction you were going before. If you keep the river to one side of you, you can always "steer" on an angle toward it to reestablish contact with the river if you need to do so.

3. Finally, don't let your frustrations undermine you. Be patient. Remember, adjusting your direction of travel along winding jungle rivers is really a process of trial and error and your main source of sustaining life—water—will always be nearby.

How to Survive Quicksand

First of all, don't panic. Despite what you might have seen in movies, it's impossible for a person to become completely submerged because the human body is less dense than quicksand—which is a mixture of sand, clay, and saltwater. The worst that can happen is that you'll sink in to just above your waist.

When people die in quicksand, they don't suffocate. Instead their feet become wedged in the densely packed sand at the bottom, and they die of thirst or starvation, or, if they're near a coast, they drown in high tide.

If you fall into quicksand (which is usually found around marshes and tidal pools), resist the natural instinct to kick your way out. All that does is separate the sand from the water, forming a very dense layer of sediment at bottom where your feet are. Instead, stay calm and lean back so you get as much of your body surface on the water as possible. Once you start to float, move your feet—not in a thrashing motion—but in small circles. You want to push more water down into the thick sediment where your feet are. As you get more of your body on the water's surface, you should be able to float free and paddle your way to solid ground.

Amazon Survival

—As Told to Andrew Taber

Two Frenchmen traveled to the Amazon to hike an approximately seventy-eight–mile section of virgin forest that was supposed to take them from their drop point on the Approuague River to Saül, an isolated former mining town at the geographical center of French Guiana.

They brought with them enough food for the eleven days they had budgeted for the trek, along with a compass, a machete, a sixty-square-foot tarp, and two hammocks.

Some days went smoothly, and they seemed to be on track. But on other days it would take several hours of hacking through vines to hike just one mile. On the morning of the twelfth day, they realized they were in trouble. Saül is in a small valley, but the route ahead of them kept rising. Exhausted and out of food, they had no idea where they were in terms of their destination. Were they two days from Saül? Two weeks?

They figured a search would be organized to find them, so they decided to stay put. They used the tarp to create a roof, and divided the tasks. One man was in charge of food, the second tended the fire. Since they had only one lighter, they kept the fire burning constantly.

Tapping into their survival instincts, they became very primal. Since it

⩔ **Dense jungle in northern Bali**

was the rainy season, water wasn't a problem. If they thought a plant was edible, one of the two Frenchman would try it. If he was okay the next day, they'd both eat the plant. They also ate beetles, bugs, and large, hairy mygale spiders (tarantulas), which they cooked until the venom burned off.

Still, they were starving. So they chewed just to chew. Psychologically, they felt it helped.

Occasionally, they'd hear helicopters, which they couldn't see through the thick canopy. Nor could the helicopters see them. Eventually the helicopters stopped passing and they started to panic.

They learned later that rescue missions did stop forty days after they had started their trek.

Around that same time, they decided that they weren't going to be found and they abandoned their camp. Using the stars, they calculated which way was west, toward Saül, and started walking. A week later, they caught a seven-pound turtle. Since it was their first meat in five weeks, they ate absolutely everything—skin, claws, scales. They even heated the blood over the fire and drank it. They claimed later that it tasted fantastic.

The following day, one of the Frenchman caught another spider. But he put it in his mouth before he'd cooked off all the venom. By the time he spit the spider out, pain engulfed the entire left side of his tongue and his lips turned numb. He tried to keep going in spite of the excruciating pain, but he was too weak. By this time he'd lost fifty-seven pounds and his partner had lost thirty-seven pounds.

The healthier of the two made a final trek into the jungle in search of help. A day and a half later, a search and rescue (SAR) helicopter came and hovered above the treetops. A SAR gendarme rappelled 150 feet to the ground and took the frail, sick Frenchman in his arms. They soon tracked down the other survivor and after fifty-one days in the jungle, their ordeal was over.

The two men learned later that they had walked seventy-five miles, but had stopped two and a half miles from Saül. They were furious with themselves.

And they had also been foolish. The Frenchmen had gone into the jungle with no local guide and had planned to travel seven miles a day. Anyone with jungle experience will tell you that seven miles a day in a virgin forest is ambitious.

The two Frenchman were also badly equipped, with only one lighter between them and no communications kit. Had they been carrying an Iridium 9505A satellite phone, they could have been tracked online. They should have also carried flares or tethered location-maker balloons that weigh about four pounds and have a chemical that inflates when mixed with water.

A simple fishing kit would have kept them supplied with readily available, protein-rich food.

In the end it was their will to survive that got them through.

4

MOUNTAIN AND ARCTIC SURVIVAL

"It is not the mountains we conquer,

it is ourselves."

—*Sir Edmund Hillary*

⌃ **Robert Edwin Peary, Sr. , an American explorer who claimed to have led the first expedition, on April 6, 1909, to reach the geographic North Pole Peary's claim was widely credited for most of the 20th century, though it was criticized even in its own day and is today widely doubted.**

⌃ **Peary with dogs on deck of *Roosevelt***

Being at high altitudes under adverse condition creates unique issues for all aspects of survival, from movement to health to dealing with injuries.

Before you set out, familiarize yourself with some of the unique challenges you might encounter:

Mountain Injuries and Illnesses

The best way to deal with injuries and sicknesses is to take measures to prevent them from happening in the first place. But it's important to treat any injury or sickness that occurs as soon as possible to prevent it from worsening. The knowledge of signs and symptoms are critical in maintaining health.

Hypothermia

Hypothermia occurs when an individual's body temperature lowers at a rate faster than it can produce heat. The initial symptom is shivering. Hypothermia begins when the body's core temperature falls to about ninety-six degrees Fahrenheit. When the core temperature reaches ninety-five to ninety degrees Fahrenheit, sluggish thinking, irrational reasoning, and a false feeling of warmth may occur. Core temperatures of ninety to eight-six degrees Fahrenheit and below result in muscle rigidity, unconsciousness, and barely detectable signs of life. If the victim's core temperature falls below seventy-seven degrees Fahrenheit, death is almost certain.

To treat hypothermia, rewarm the entire body. If there are means available, rewarm the person by first immersing the trunk area only in warm water of one 100 to 110 ten degrees Fahrenheit. Never rewarm the entire body in a warm bath, as this can result in cardiac arrest and rewarming shock.

One of the quickest ways to get heat to a person's inner core is to administer warm water enemas. But such action may not be possible in a survival situation. Another method is to wrap the victim in a warmed sleeping bag with another person who is already warm; both should be naked.

If the individual suffering from hypothermia is still conscious, feed him hot, sweetened fluids. One of the best sources of calories is honey or dextrose; if unavailable, use sugar, cocoa, or a similar soluble sweetener. Never force an unconscious person to drink.

The two dangers of treating hypothermia are:

1. Rewarming too rapidly. This can cause circulatory problems, resulting in heart failure.

2. "After drop." After drop is the sharp body core temperature drop that occurs when you remove the victim from the warm water. Warming the core area and stimulating peripheral circulation will lessen the effects of after drop. Immerse the individual's torso in a warm bath, if possible.

Frostbite

Frostbite is a constant danger to anyone exposed to subzero temperatures. Because it isn't painful, frostbite often goes unnoticed. Therefore, when operating in subzero temperatures, you need to frequently examine your face, hands, and feet.

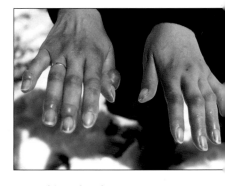

▲ **Frostbitten hands**

Frostbite can occur on exposed skin:

- within thirty minutes below minus 22 degrees F
- within one minute below minus 60 degrees F
- within thirty seconds below minus 74 degrees F

Feet, hands, ears, and exposed facial areas are particularly vulnerable. Skin that becomes dullish white in pallor is an indicator of light frostbite. Deep frostbite extends to a depth below the skin.

The best way to prevent frostbite when you are with others is to use the buddy system. Check your buddy's face often and make sure that he checks yours. If you are alone, periodically cover your nose and lower part of your face with your mittened hand.

The following pointers will aid you in keeping warm and preventing frostbite when it is extremely cold or when you have less than adequate clothing:

- **Face**. Maintain circulation by twitching and wrinkling the skin on your face. Warm with your hands.
- **Ears**. Wiggle and move your ears if you can. Warm with your hands.
- **Hands**. Move your hands inside your gloves. Warm by placing your hands close to your body.
- **Feet**. Move your feet and wiggle your toes inside your boots.

Loss of feeling in your hands and feet is an indicator of frostbite. If you have lost feeling for only a short time, the frostbite is probably light. Otherwise, assume the frostbite is deep.

To rewarm a light frostbite, use your hands or mittens to warm your face and ears. Place your hands under your armpits. Place your feet next to your buddy's stomach. A deep frostbite injury, if thawed and refrozen, will cause severe damage.

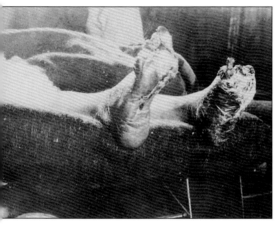

≈ **Case of trench foot suffered by unidentified soldier during World War I**

An old remedy for frostbite called for rubbing ice or snow into the affected part, but this will only lower the temperature even more and risk removing the outer layer of skin. As with most types of cold injuries, the best solution is to warm the affected part gradually. Don't rub the spot. Even the gentlest massage can do a great deal of harm, as mentioned earlier. If frostbite appears on your face, warm it by pressing your warm fingers against it. If a wrist is frozen, warm it by grasping it with the other hand. Frozen hands and fingers can be thawed by holding them against your chest or under your armpits inside your clothes.

Frozen feet are particularly serious because there's a high risk of losing toes. Keep your feet from freezing by using warm, insulated footwear. If you suspect your feet are frostbitten, take care of them immediately. Change to warm, dry footgear if you can, or wrap them in cloth until they thaw. Warm them gradually. Don't put them close to a heater or a fire. A burning sensation will follow the warming and thawing of a frozen part, and can be extremely painful. After frostbite, there may be blistering and peeling of the skin, as in sunburn, because the extreme cold has "burned" your skin.

Trench Foot and Immersion Foot

Both conditions can result from hours or days of exposure to wet or damp conditions at a temperature just above freezing. Initial symptoms are a sensation of pins and needles, tingling, numbness, and then pain. The skin will appear wet, soggy, white, and shriveled.

If untreated, the skin will take on a red and then a bluish or black discoloration. The feet will become cold, swollen, and have a waxy appearance. Walking will be difficult as the feet will feel heavy and numb. The nerves and muscles sustain the main damage, but gangrene can occur. In extreme cases, the flesh dies and it may become necessary to have the foot or leg amputated

The best prevention is to keep your feet dry. Carry extra socks with you in a waterproof packet. Dry wet socks against your torso (back or chest). Wash your feet and put on dry socks daily.

Dehydration

When you're bundled up in many layers of clothing during cold weather, you may be unaware that you are losing body moisture. Heavy clothing absorbs the moisture that evaporates in the air. It's necessary to drink water to replace this loss of fluid. Your need for water is as great in a cold environment as it is in a warm environment.

One way to check to see if you're becoming dehydrated is to observe the color of your urine on snow. If your urine makes the snow dark yellow, you're becoming dehydrated and need to replace body fluids. If it makes the snow light yellow to no color, your body fluids have a more normal balance.

Cold Diuresis

Exposure to cold actually increases urine output, which means that those body fluids need to be replaced.

Sunburn

Exposure to the sun results in sunburn more quickly at high altitudes than at low altitudes. When in a mountain environment, the sun's rays also reflect at all angles from snow, ice, and water, hitting sensitive areas of skin—lips, nostrils, and eyelids. Always apply sunburn cream or lip salve, if available, to your face when in the sun.

Snow Blindness

This can result from the reflection of the sun's ultraviolet rays off a snow-covered area. The symptoms of snow blindness are a sensation of grit in the eyes, pain in and over the eyes that increases with eyeball movement, red and teary eyes, and a headache that intensifies with continued exposure to light. Prolonged exposure can result in permanent eye damage. To treat snow blindness, bandage your eyes until the symptoms disappear.

Snow blindness can be prevented by wearing sunglasses or, better yet, glacier glasses. If you don't have either, improvise. Cut slits in a piece of cardboard, thin wood, tree bark, or other available material. By placing soot under your eyes, you will help reduce shine and glare.

Even brief exposure to the sun on a relatively overcast day can result in snow blindness, so take precautions.

≫ **Glacier glasses**

Constipation

It is important to relieve yourself when needed, even in extremely cold conditions. Eating dehydrated foods, drinking too little liquid, and irregular eating habits can cause you to become constipated. Although not disabling, constipation can cause discomfort. Increase your fluid intake to at least two liters above your normal two- to three-liter daily intake and, if available, eat fruit and other foods that will loosen the stool.

↟ **Black Canyon, Colorado**

Stages of Altitude Sickness

There are a variety of illnesses that can afflict poorly acclimated individuals, usually occurring within the first several days of ascending too quickly to altitudes greater than 8,000 feet. These are caused by low atmospheric pressure, ascending too quickly, high activity levels, dehydration, excessive consumption of alcohol, poor diet, and/or the use of over-the-counter sleeping medications.

1. **Mild Acute Mountain Sickness (AMS)** Individuals with AMS have headaches, shortness of breath when exercising, loss of appetite, insomnia, weariness, and fatigue. (Similar to an alcohol hangover.)

    Treatment: Wait for improvement before ascending further! Take either aspirin or acetaminophen to treat headaches. The prescription drug acetazolamide (Diamox) may reduce the incidence and the severity of AMS. Increase water consumption and eat more carbohydrates. Symptoms will usually clear up within twenty-four to forty-eight hours. Those experiencing mild AMS should consider it a warning and take time to acclimatize before continuing. Those that do not acclimatize well should descend to lower altitudes.

⌄ **Franz Josef Glacier, New Zealand**

2. **Moderate Acute Mountain Sickness.** The symptoms of mild AMS have progressed to the point that the victim is very uncomfortable. Severe headaches that are only partially relieved with aspirin (if at all), weakness, weariness, fatigue, nausea, breathlessness at rest, and lack of coordination are common symptoms.

    Treatment: Persons with moderate AMS must **stop ascending** and, if the condition does not improve, must **descend** to lower altitudes. Failing to recognize what is happening and not descending quickly can result in a life threatening medical emergency—either High Altitude Pulmonary Edema (HAPE) or High Altitude Cerebral Edema (HACE), which occur within hours and can result in the death of the victim.

**Norway—U.S. Navy SEALs and German forces free-fall »
parachute onto a frozen a lake in northern Norway during
exercise cold response. Cold response is a Norwegian-
sponsored multinational invitational exercise, with more than
9,000 military personnel from fourteen countries focused on
cold weather maritime and amphibious operations.**

3. **High Altitude Pulmonary Edema (HAPE).** HAPE progresses to life-threatening seriousness in only a matter of hours. Early signs include marked breathlessness on exertion, breathlessness at rest, decreased exercise capacity, increased respiratory and heart rate. In moderate to severe HAPE, there is marked weakness and fatigue, bluish discoloration of the skin, a dry raspy cough, and gurgling sounds in the chest. As HAPE worsens, a productive cough develops.

   Treatment: Immediate descent to lower altitudes is essential. Descend 2,000 to 4,000 feet—get below 8,000 feet if possible. Keep the victim warm. Exert the patient as little as possible. Advanced treatment may consist of administering acetazolamide in mild cases of HAPE and nifedipine in moderate to severe cases.

How to Prevent HAPE

— Climb slowly—the faster the rate of ascent, the more likely it is that symptoms will occur.

— Lay over at intermediate altitudes before ascending to the final altitude.

— Avoid overexertion.

— Increase consumption of water.

— Avoid alcohol.

— Don't use over-the-counter sleeping aids.

— Eat more carbohydrates.

— Sleep at lower altitudes.

— If you know you are susceptible, consult your doctor about appropriate medications.

4. **High Altitude Cerebral Edema (HACE).** HACE is the result of swelling of brain tissue from fluid leakage and almost always begins as acute mountain sickness (AMS). Symptoms usually include those of

≈ **Author and his climbing partner at 19,500 feet descending a route in the Andes** ≈

≈ **Denali—snow wall built around tent for wind barrier**

AMS (nausea/vomiting, insomnia, weakness, and/or dizziness) plus headache, loss of coordination (ataxia), and decreasing levels of consciousness including disorientation, loss of memory, hallucinations, irrational behavior, and coma.

Treatment: Oxygen administration and medications (dexamethasone) may temporarily alleviate symptoms and facilitate descent, which is the necessary life-saving measure. Hyperbaric bags are highly effective in conjunction with dexamethasone and are relatively inexpensive and lightweight (fifteen pounds). Evacuated patients should go to a medical facility for follow-up treatment.

Mountain Weather

Mountain weather can change quickly and dramatically. And severe bad weather can strike unexpectedly. Suddenly you can become blinded by snow and so numb with cold that you can lose feeling in your outer extremities. The best way to deal with it is to prepare ahead of time by packing the right gear and learning how to predict and deal with severe weather conditions.

WindChill Factor

The windchill factor is the temperature you feel on your exposed skin because the wind affects how quickly we lose heat from our bodies. In arctic environments, where altitude and wind speed can be high combined with a low air temperature, windchill can be very dangerous.

The importance of the windchill index is as an indicator of how to dress properly for winter weather. In dressing for cold weather, an important factor to remember is that entrapped insulating air warmed by body heat is the best protection against the cold. Consequently, wear loose-fitting, lightweight, warm clothing in several layers. Outer garments should be tightly woven, water-repellant, and hooded. Mittens that are snug at the wrist are better protection than fingered gloves.

To use the following chart, find the approximate temperature on the top of the chart. Read down until you are opposite the appropriate wind speed. The number which appears at the intersection of the temperature and wind speed is the windchill index.

NWS Windchill Chart

Temperature (°F)

Calm	40	35	30	25	20	15	10	5	0	-5	-10	-15	-20	-25	-30	-35	-40	-45
5	36	31	25	19	13	7	1	-5	-11	-16	-22	-28	-34	-40	-46	-52	-57	-63
10	34	27	21	15	9	3	-4	-10	-16	-22	-28	-35	-41	-47	-53	-59	-66	-72
15	32	25	19	13	6	0	-7	-13	-19	-26	-32	-39	-45	-51	-58	-64	-71	-77
20	30	24	17	11	4	-2	-9	-15	-22	-29	-35	-42	-48	-55	-61	-68	-74	-81
25	29	23	16	9	3	-4	-11	-17	-24	-31	-37	-44	-51	-58	-64	-71	-78	-84
30	28	22	15	8	1	-5	-12	-19	-26	-33	-39	-46	-53	-60	-67	-73	-80	-87
35	28	21	14	7	0	-7	-14	-21	-27	-34	-41	-48	-55	-62	-69	-76	-82	-89
40	27	20	13	6	-1	-8	-15	-22	-29	-36	-43	-50	-57	-64	-71	-78	-84	-91
45	26	19	12	5	-2	-9	-16	-23	-30	-37	-44	-51	-58	-65	-72	-79	-86	-93
50	26	19	12	4	-3	-10	-17	-24	-31	-38	-45	-52	-60	-67	-74	-81	-88	-95
55	25	18	11	4	-3	-11	-18	-25	-32	-39	-46	-54	-61	-68	-75	-82	-89	-97
60	25	17	10	3	-4	-11	-19	-26	-33	-40	-48	-55	-62	-69	-76	-84	-91	-98

Wind (mph)

Frostbite Times ▢ 30 minutes ▢ 10 minutes ▢ 5 minutes

Wind Chill (°F) = $35.74 + 0.6215T - 35.75(V^{0.16}) + 0.4275T(V^{0.16})$

Where, T= Air Temperature (°F) V= Wind Speed (mph)

Effective 11/01/01

⌃ **Chart and facts courtesy of National Weather Service**

⌃ Avalanche ⌄

Avalanches

Avalanches are always a big threat in the mountains. They kill approximately fifty-eight skiers every year in North America alone. When traveling to an avalanche risk area, carry a beacon that, when activated, will transmit a signal that the rescue services will follow if you get lost or buried in the snow.

The key to avoiding avalanches is to read the snow. Use a ski pole to test the snow to see if it's compacted or in layers. If it's consistent when you push in the snow, it's probably okay. If it suddenly drops off, that indicates it's in layers and dangerous.

What to do if You're Hit by an Avalanche

Immediately after you're swept away, fight like hell to stay on the surface of the slide. But once you're engulfed, most avalanche experts now agree that "swimming" won't help you rise to the surface. Worse, after the snow slams to a stop, all that arm waving will have left you without an air pocket. The safer bet: Keep your hands near your face while you're tumbling. And, if possible, try to stick one hand above the surface. (Your chances of being found increase exponentially if a part of you is visible.)

An Avalanche Story

The first avalanche hit at midnight. Seconds later the walls of two ultralight tents that four American climbers were sleeping in collapsed. They awoke to ice and snow squeezing them in the darkness.

⩘ **Mount St. Helens looking south directly toward the bulge. Numerous rock avalanches were triggered by earthquakes.**

It was late September, and the four climbers were deep in the Kumaon region of the Indian Himalayas to climb 22,510-foot Nanda Kot.

During their first attempt on the peak, they had been hit by a severe north-blowing storm that killed more than a dozen people.

Temperatures dropped to below zero, and heavy clouds dumped more than six inches of snow per hour. They took shelter by chiseling their tents inside a crevasse that sloped downward into a seemingly bottomless pit. The orientation of the opening protected them from the full force of the avalanche that followed, but the snow still poured in fast and deep.

As the cement-like mass pushed them farther into the crevasse, the four American climbers fought to keep an air space in front of their mouths while tearing through the tent and pulling up frantically. When the slope finally settled, one of the three male climbers ended up with his head and one arm above the surface. He immediately started digging and found the sole female climber trapped near his knees. He burrowed through the snow and ice until her hand grabbed his. Upon hearing the roar of the approaching avalanche, a second male climber grabbed an anchor they had set in the crevasse wall. He caught the other male climber's arm just as their tent was crushed. Miraculously, all four climbers survived.

Using the light from a headlamp, they began digging for their boots in the ten feet of debris. They knew they needed them if they wanted to make it off the mountain. After six hours of digging, they had dislodged everyone's boots, their ice tools, a stove, and four canisters of fuel. As the sun started to rise, they cooked some dehydrated noodle soup and were sitting on their shredded tents when a second avalanche hit. The four climbers braced. It was 6:30 AM, but suddenly it turned dark again.

⩘ **K2**

One headlamp popped on, then another. The climbers saw that they were trapped in a ten-by-twenty-foot space. Two of them dug a fifteen-foot-long tunnel through the wall and poked their heads out into a raging storm. They knew immediately that there was no way they could have survived out there. So they decided to stay in the chamber until they either ran out of fuel or the storm surrendered. They figured that their chances were thin. The crevasse's unstable wall looked like it might collapse under the weight of the avalanche debris and new snow. And they were unroped. A fall into the abyss below would be fatal.

On the second day, the four climbers came up with ways to distract themselves—building chairs out of snow and playing twenty-one questions. But by the third day, they couldn't escape the reality of their situation: They were going to die there if they didn't get out soon. Luckily for them, the storm lifted on the fourth day and the slopes began to stabilize. They tunneled out and spent the next twelve hours rappelling down 6,000 feet.

The crevasse probably saved their lives. Most climbers are taught to avoid crevasses in a potential avalanche situation since they're considered terrain traps. These four climbers were extremely lucky.

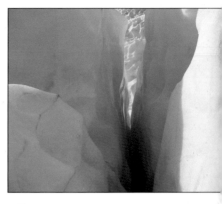

≈ **Crevasse**

Steps to Take if You Find yourself Abandoned on a Snowy Mountain:

- Keep your extremities as warm as possible and watch for signs of frostbite—waxy, red-black skin.
- Glaciers are a good landmark. If you follow them down, they'll lead you out of the mountains.
- If you need to find your bearings on a sunny day, you can find north, east, south, and west by using the shadow and stick method. Find a stick, insert it into the ground, and you'll see that it casts a nice shadow. Mark where the end of the shadow is and leave the stick for fifteen minutes. Mark the next point and that will create an east-west line.
- Often dew will collect on leaves and pine needles overnight, so if you're desperately in need of water, this can be a lifesaver.
- If you are lucky enough to successfully hunt food to cook, remember to dispose of any leftovers or else you're at risk of black bears in the area coming for your leftovers.

≈ **Glacier National Park, Montana. Crevasses near the terminus of Grinnell Glacier.**

Keizo Funatsu—Lost Overnight in a Hurricane-Force Blizzard

Keizo Funatsu was the sole Japanese member of the six-man International Trans-Antarctic Expedition of 1989-1990 that had successfully dog-sledded 3,725 miles across Antarctica. They were sixteen miles from their destination—Russia's Mirnyy base—when Keizo Funatsu vanished in a raging blizzard.

He'd gone outside at four-thirty in the afternoon to feed the sled dogs. At that time snow had stated falling and it was about minus twenty-five degrees F. Within minutes the wind picked up dramatically, and visibility became problematic. Keizo, who was wearing only Gore-Tex boots, a wind parka, and wind pants, long underwear, and wool socks, couldn't find his second ski, even though he'd planted his ski poles upright in the snow to serve as guideposts.

He waited ten minutes for the visibility to improve without any luck. When he tried to move back toward his first ski, he found it almost impossible to walk straight ahead into the wind. He tried moving left, then right, and realized that because of the lack of visibility, he wouldn't be able to find anything even though his skis were, at most, a few feet away.

He tried shouting, "I'm here. I'm here. Come on." Frustrated by the fact that he wasn't far from the camp, he tried pushing through the wind. Initially the cold didn't bother him because the adrenaline in his body was keeping him warm. But Keizo was worried about his toes, which were getting cold.

He spotted some dog excrement and a faint sled trail of a sled, which he tried to follow. It quickly disappeared. The sled trail meant that Keizo was positioned behind the camp, so he stopped.

Using a pair of pliers he carried in his pocket to help unfreeze dog collars and fix broken ones, Keizo started to dig into the icy surface and scooped out a shallow ditch about two and a half feet deep. He put his feet in and allowed the storm to bury him just as he'd watched the sled dogs do to protect them from the elements.

The blowing snow covered him in seconds. Now, he was completely covered and breathing through an air hole close to his body. Unfortunately he wasn't wearing much and the snow, which was heavy, pressed down on

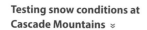

Testing snow conditions at Cascade Mountains ⌄

his wind jacket, so his clothes touched his skin. That left no room for a layer of warm air to form.

As time wore on, Keizo realized that he'd probably have to spend the night in the snow ditch since the storm hadn't let up and there was probably zero chance of his teammates finding him in the dark.

He was cold and worried about losing energy. He didn't know if he should lie still or move around to try to warm his body. He kicked his feet to keep the blood circulating. Every twenty or thirty minutes he pulled himself out of the snow ditch, jumped up and down, and rubbed his arms. At first, he tried shouting but realized that he couldn't be heard through the raging wind.

The wind was so strong, in fact, that it pushed him away from the snow ditch. He had to crawl on his stomach to find it. Cleverly, he'd spread everything he had in his in pockets around it as guides—pliers, headband, lip cream, compass, pocket knife.

Keizo tried to think positively, saying to himself: "Very few people have had this kind of experience, lost in a blizzard. Settle down, and try and enjoy this." With the snow and quiet covering him, he felt as though he was back in his mother's womb. His life seemed infinitesimally small in comparison to the vastness of Antarctica.

Keizo started to feel shame. Alone in the snow ditch, he thought about the responsibility he had to the rest of the team and the trouble he would cause them if he died sixteen miles away from their destination—Mirnyy.

The sky above him started to lighten around five o'clock in the morning. Keizo climbed out of the snow ditch and tried to locate the sled trail again, but fresh snow covered everything. He was concerned about his feet, which weren't cold, but felt like they were swelling. This was a sign of frostbite. He was afraid to take his socks off and rub them because then his feet would get wet.

Returning to the snow ditch, Keizo thought he heard somebody yelling faintly outside, "Keizo! Keizo!" The storm hadn't let up, so he figured it must be the wind playing tricks on him. He'd been hearing things all night, and had learned how the wind could sound very much like a human voice.

Still, he jumped out of the ditch to look. He heard the voice again and yelled back, "I'm here. I'm here!"

He couldn't see anything through the falling snow. Finally he heard a voice close by and screamed, "I'm here. I'm here!" He made out team leader Will Steger's silhouette through the blanket of white. Leaving his pliers and everything else behind, he ran toward him.

Keizo saw that the rest of the six-member team was out in the blizzard looking for him—walking in a giant circle, linked by a three-hundred-forty-foot rope. They were thrilled to find him. Some of them had tears in their eyes. He cried, too.

⌃ **Maintain proficiency in rescue training**

Dr. Beck Weathers—Surviving Mt. Everest

Plagued by depression, a turbulent home life, and seeking adventure, Dr. Seaborn Beck Weathers of Texas set off in 1996 with nine mountain climbers to tackle mighty Everest.

Dr. Weather's trouble on the mountain began when he lost his vision at about 27,500 feet, caused by the effects of high altitude on his eyes, which had been altered by radial keratotomy surgery. After discussing this disability

⌃ **Mt. Everest**

with his guide, Rob Hall, Weathers decided to remain on the balcony of the summit ridge until the other climbers descended back to him.

But that afternoon a ferocious blizzard blew in from the south. Hall and the other climbers became trapped and were unable to descend. He and five others died.

Weathers, meanwhile, spent the night in an open bivouac in subzero temperatures, blind, and with both hands and his face exposed. When the surviving climbers reached him, they said that his frozen hands and nose looked and felt as though they were made out of porcelain. Seeing that he was in a deep hypothermic coma, they left him for dead.

Miraculously, Weathers survived another freezing night alone in a tent unable to drink, eat, or keep himself covered. His cries for help couldn't be heard through the blizzard even though his companions were bivouacked in another tent nearby. They were shocked to find him alive and coherent the following day.

Weathers recovered enough to walk unassisted to the nearby camp. From there he continued on his frozen feet to a lower camp, where he was rescued by helicopter.

Weathers had his right arm amputated halfway between the elbow and wrist. All four fingers and the thumb on his left hand were removed. His nose was amputated and reconstructed with tissue from his ear and forehead. And, he lost parts of both feet.

Despite these disabilities, Weathers returned to working full-time as an anatomic pathologist. He said that his ordeal made him a better, happier, more spiritual person.

Survival in the Andes

On October 12, 1972, a chartered military plane carrying an amateur rugby team (known as the Old Christians) and some members of their families left Montevideo, Uruguay, for Santiago, Chile. Because of bad weather conditions over the Andes, the plane was forced to land in Mendoza, Argentina, where they waited two days for the weather to improve.

The mood aboard the plane was festive as the group finally took off again. But as they approached Santiago, they hit more bad weather. The inexperienced pilots of the Fairchild F-227 lost control of the plane causing it to crash into the side of a mountain and lose both wings and the tail. The aircraft then slid on its belly into a step valley surrounded by snow and mountains.

Because the roof the aircraft had been painted white, it was almost impossible to spot by air rescue. Chilean, Argentine, and Uruguayan military and civilian teams searched for more than a week but couldn't find any sign of the wrecked aircraft.

Thirty-two of the forty-five passengers survived the initial crash. Within days, without adequate medical attention, the most severely injured died in the harsh subzero conditions.

≈ **Andes Mountains, Peru**

None of the survivors were prepared to face the severe cold, and their provisions were limited to some alcohol, candy, and a few miscellaneous items like crackers and jam. As the survivors grew weaker and weaker, they realized that the only way to survive was to eat the flesh of their dead colleagues.

On the seventeenth day they were hit by an avalanche, and thirteen more people died. The remaining survivors realized that the only way to make it out alive was to climb out of the steep valley and find help.

After ten days of trekking through some of the highest mountains in the world, Nando Parrado and Roberto Canessa came across a Chilean peasant who was tending his animals in a remote valley in the Andes. Initially, the peasant ignored them when they tried to get his attention, fearing they might be terrorists. But Parrado and Canessa persisted, and threw a piece of paper and a pen wrapped in a handkerchief to him. The paper read, "I come from a plane that fell in the mountains. I am Uruguayan . . ."

On Thursday, December 21st—seventy days after the crash—sixteen survivors were finally led out of the remote valley and taken back to civilization in what became known as "the Christmas Miracle."

The ordeal was especially tragic for twenty-one-year-old Fernando (Nando) Parrado, who had to bury his mother and sister, who had accompanied him on the trip, in an arid, freezing glacier.

Nando Parrado said later that he learned some important lessons about survival:

≈ **Fernando Parrado and Roberto Canessa next to the cattle driver who discovered them.**

1. Making quick decisions is a great virtue. Those who don't decide, die. "If I make the wrong decision I have time to correct," Parrado explained. "It's far better to decide and make mistakes, than not making decisions, because there is always time to go back."

≈ **Author with his climbing partner descending Cotopaxi (17,750 feet) in an attempt to beat the oncoming storm.**

≈ **Author climbing in the Andes Mountains.**

≈ **Author and his climbing partner trying to warm up in an ice cave**

2. Although democratic decision-making is good at certain moments, sometimes a leader has to step forward to make decisions because it's not always easy for a group to reach a consensus.

3. Leaders aren't born; they develop their skills with their actions. In other words, leaders are those who achieve good results.

4. Obsessive focus on objectives and results will keep you alive.

5. Creativity is needed to find solutions. Parrado explained how he and others made plates out of pieces of aluminum that were used to melt the snow into potable water. "Snow was more important than food," he said, "because the human body dehydrates five times faster at 11,500 feet than it does at sea level."

6. Define which things are important and which ones are not. He explained, "A hundred percent of the people who were trapped with me in the Andes wanted to go back to their families, not to their contracts, studies, or money. In fact, we burned all the money on the plane ($7,000) to keep warm."

Arctic Survival

Arctic conditions extend into Alaska, Canada, Greenland, Iceland, and much of northern Scandinavia and Russia. The Arctic has no land mass but is a floating sheet of permanently frozen ice. The Antarctic, on the other hand, consists of a large land mass covered by a permanent ice layer, which in some places is more than 1,000 feet thick.

Both regions have two seasons—long winters featuring days with twenty-four hours of darkness, and short summers with days that can have twenty-four hours of daylight. Summer temperatures in the Arctic can rise to sixty-five F, except on glaciers and frozen seas. During winter they fall to as low as minus eighty-one F and are never above freezing.

Antarctic temperatures can dip even lower. The combination of extremely cold air temperatures, high altitudes, and high winds produce the most hostile environment known to man. The average altitude in Antarctica is 2,300 meters (7,540 feet), with the highest point at around 4,000 meters (13,100 feet). Antarctic winds as high as 177 kilometers (110 miles per hour) have been recorded. Winter winds sometimes reach hurricane force and can whip snow thirty meters (100 feet) into the air, giving the impression of a blizzard even when it isn't snowing.

The polar regions of both areas are devoid of large flora species. On the other hand, aquatic animals such as the walrus and seal (as well as whales and birds, including the penguin found in the Antarctic) have adapted to the extremely low temperatures. The North Pole is home to the world's largest land carnivore, the polar bear, which is one of the few animals that will actively hunt humans.

Located south of the North Pole are large expanses of tundra, where plant life is limited because roots can't penetrate the frozen ground. The only plants that grow are reindeer moss, which is low, bushy, coral-like lichen that grows on the ground common throughout northern Canada, and lichens, which must be thoroughly boiled or soaked in water for several hours before being eaten.

South of the tundra is a vast area of coniferous forest, called the boreal forest, that extends to the temperate zone. In Russia, this forest is known as Taiga and covers the areas along the Siberian rivers northwards into the Arctic Circle. Coniferous forests, which are also found in northern Canada, contain a great deal of wildlife including bears, elk, reindeer, lynx, wolf, sable, wolverine, and lemmings.

Hunting

Local tribes, people such as the Inuit of northern Canada and Alaska and the Sami of northern Finland, have survived by being skillful hunters. The Sami also domesticate reindeer, whose milk is nutritious but low in yield, and whose venison has a rich flavor. It's also fairly lean, which isn't ideal in the cold temperatures where fat is needed to help the body generate heat.

Musk ox has a strong flavor but is very high in fat. The Arctic hare is very low in fat, so a diet consisting mostly of hares will not be sustaining. Various parts of caribou can be eaten, especially meat from the head, neck, shoulder, hindquarters, heart, liver, brisket, ribs, backbone and pelvis. In the unlikely event that you kill a polar bear, you'll find that the meat is tough and stringy. Some claim that it's more tender eaten raw. Polar bear livers are poisonous, however, so don't eat them. Seals are also edible.

Certain arctic birds—such as ducks, geese, and swans—have a high fat content. During summer, many of these water birds are unable to fly for two to three weeks while they're molting and can make easy prey. Snow partridge, gull, and tern colonies can supply a good amount of eggs but are usually hard to get to because they are usually located on small islands or high on cliffs.

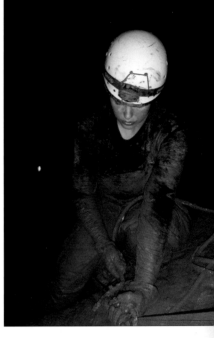

≈ **Checking climbing gear in a muddy cavern**

« Cascade Mountains—using sled to protect gear from snow.

« Alaska—author and his team breaking down camp

« A tent can be quickly buried in a snow storm. Be sure to get up every hour during a storm to remove snow from tent and tent entrance.

Environmental Hazards

Arctic areas are some of the most inhospitable areas on the planet. In tundra areas, expect travel to be difficult because of deep snow and dense forests. The locals often use the frozen rivers as highways, as these are generally wide and free of obstructions. Of course, in spring these rivers can rapidly thaw and become extremely dangerous. Also be aware that it's rare for large rivers to freeze completely, so don't be alarmed if you're walking on a foot of ice and can still hear water running under your feet.

If you're lost, it's usually best to follow rivers downstream, as this will usually lead to some form of civilization. The exception to this general rule is in Siberia, where many rivers flow north and away from any inhabited areas.

Wind Chill and Breathing

In addition to the extremely low temperatures, upland areas in particular can be subject to very strong arctic winds. Inhaling cold air can actually burn the lungs and cause hemorrhaging. To reduce the risk of this happening, learn to control your breathing by exhaling completely but inhaling slowly and shallowly at first until your lungs build up some resistance to the cold. Controlling your breathing can also help you focus and remain calm during stressful conditions. Remember that a rapid intake of subzero air can cause considerable damage to your lungs.

5

DESERT SURVIVAL

"A desert is a place without expectation."

—*Nadine Gordimer*

≈ **Desert outback**

≈ **Dry desert sand**

Deserts are one of the most hostile human environments on the planet. The extreme heat and aridness make desert survival extremely challenging. Add other factors, both physical and psychological, and conditions can be unbearable.

A desert is an area that receives almost no precipitation (rain, snow, moisture), with an annual precipitation of no greater than twenty-five centimeters per year. Surprisingly, the world's largest deserts are Antarctica and the Arctic in that order, followed in turn by the best known hot desert, the Sahara.

The world's ten largest deserts, in size order:

- Antarctic
- Arctic
- Sahara
- Arabian
- Gobi
- Kalahari
- Patagonian
- Great Victoria Desert
- Syrian Desert

Natural Hazards

Most deserts are featureless (or have repeating patterns of sand dunes), making terrain navigation difficult, and have virtually no vegetation. The majority of desert animals are nocturnal and remain hidden underground during the day to conserve water and regulate body temperature. These include coyotes, kangaroo rats, and jackrabbits.

A wide variety of insects and reptiles can also be found, but many of these are poisonous either to eat or have poisonous bites or stings. These include scorpions, which have a tendency to lurk in footwear. So it's important to make sure that footwear is hung upside down or checked in the morning before putting on.

Heat exhaustion is the most common problem caused by loss of salt and fluid. The signs include weakness, headaches, pale clammy skin, and mental confusion. Heat stroke has the same causes but can be fatal if not treated quickly. Symptoms normally include hot dry skin (unlike heat exhaustion), headache, vomiting, a fast pulse rate, and a confused mental state, which usually precedes unconsciousness and death. Heat cramps are a result of loss of salt and are muscle cramps that start in the limbs and can spread throughout the body, eventually preventing any physical activity.

Sandstorms can last days and make navigation and travel extremely difficult. They also can clog vehicle filters and get into equipment, doing damage to everything from vehicles to communications equipment. Keep a scarf or a piece of cloth over your mouth and nose, and wear goggles or at least sunglasses during a sandstorm.

The best advice for most people lost in the desert is to get into the shade immediately and wait to be rescued. Obviously this rule does not apply during desert warfare training or during desert operations.

Heat

Desert heat can reach over 100 degrees F and can incapacitate and kill you in a matter of hours. In the night, the temperatures can drop to near freezing. Do everything you can to keep cool and stay out of the sun.

Look for an outcrop of rock, the shady side of a gully or streambed, or any shade you can find. Keep in mind that you're searching for an emergency shelter for a few hours, perhaps, not a long-term one. You'll have time to construct a better shelter after the sun goes down. The critical thing is to get out of the sun and into the shade as soon as possible.

Keep your whole body covered if you can. Keep your sleeves rolled down and never remove your boots, socks, or any piece of clothing while in the direct sunlight. Additionally, cover the back of your neck to protect it from the sun. If you're wearing a T-shirt, remove it and use it as a scarf. Push one end of the shirt up under your cap, and allow the other end to hang over the neck. The most important part of your body to keep cool is your head. Always wear a hat and, if for some reason you do not have one, make a headdress out of light-colored material. Regardless of the technique you use, remember to keep your neck covered at all times. It reduces water loss through sweating and it also prevents sunburn.

Most desert survival experts agree that in the 120 degree-plus heat of a desert, if you rest and do nothing, you may live for a couple of days. If you go moving into the desert, you'll most likely cover less than five miles. If you wait until after the sun sets, you may be able to cover up to twenty-five miles or so. So, if the sun is out, stop and rest. Go no farther and seek shade immediately!

Once you're in the shade, continue to try to establish communications with your support assets. They should know where you are, how long you intend to be there, and the exact time and date you planned to return. Be

≫ **Sunburn**

sure to communicate any changes to your plans. There's nothing more frustrating to rescue teams than to be searching for someone who is not where he's supposed to be.

While in the shade, inventory the equipment and the survival kit you have on hand.

Heat Acclimation

It takes approximately two weeks to fully acclimate to hot environments. Your body will eventually adjust to the following physiological changes:

- Sweat rate increases
- Sweat is more diluted
- Heart rate decreases
- Body temperature increases

Your body's core temperature rises during exercise and hard work in a hot environment. This causes your heart rate to increase, your skin temperature to increase, and your sweat to become more profuse. And this all causes your heart to have to work harder. Your body pumps blood to the surface of your skin to cool; it is pumped to the heart and muscles. As your body heats up, it has to work harder to cool your blood and keep your muscles working.

Your body also loses fluid when you sweat, and you can easily become dehydrated. If you lose more than two percent of your body weight through dehydration, you will start losing your ability to perform physically.

It is VERY common to underestimate how much fluid you lose when you sweat. Drink fluids early and often. Don't wait until you are thirsty to take a drink.

Desert Survival Tips:

- Drink more water and fluids per hour than you think you need (one liter per hour or more as needed during hard work or exercise).
- Consume potassium-rich foods such as bananas, parsley, dried apricots, dried milk, chocolate, various nuts (especially almonds and pistachios), potatoes, bamboo shoots, avocados, soybeans, and bran, Potassium is also present in sufficient quantities in most fruits, vegetables, meat, and fish.
- Add salt to your diet to help your body store water more efficiently.
- Cover your skin with light-colored, lightweight, loose long sleeves and pants when appropriate.

≈ **Desert landscape**

- Wear a hat to keep the sun off of your head and face.
- Splash cool water on your head or wet your hat during workouts to cool your core temperature.
- Wear sunglasses to protect your eyes.
- Avoid the hottest part of the day (10–4 PM) if possible.

Desert Recommended Gear

Your mission, mode of travel, and expected length of operation will determine the equipment you will carry.

When traveling by vehicle—be it a Humvee, ATV, truck, or Jeep—you have the space, so pack gear such as shovels, picks, and heavy rope.

⌃ **Trekking in Death Valley**

Every SEAL, every experienced operator who trains and operates in the desert, will typically carry a well thought-out survival or E&E (escape and evade) kit to suit their operational and personal needs. They will learn to use it and will practice with each item before a mission.

Below is a list of equipment often found in many survival and E&E kits.

Signal mirror (can use the one inside of a Silva Ranger compass)	Windproof lighter x2	GPS	Heavy-duty aluminum Foil
Leatherman type of multipurpose tool	Medical kit	Compass	Electrolyte tables
Flint and steel	Nalgene or similar drinking container	Watch with altimeter, barometer, and compass	Tweezers
Woodman's saw	Linen thread (one yard)	Map(s)	Razor blade
Small candle	Light picture cord (snare wire)	Duct tape	Heavy plastic bags for bivy sack or water-
Small pencil	Hard candy	Bandanna	proofing shelter
Comb	Energy bars	1 square yard nylon or chiffon	Knife sharpening device
Bar of soap	MRE(s)	Local currency	Water, one gallon per
Toothbrush	Space blanket	Rain gear	person per day; five
Condoms for water storage	Fishing kit with hooks, sinkers, and some line.	Sheets of plastic or Tyvek	gallons per vehicle
Thunder whistle		Needle with large eye	Water purification tables
		Survival guide	Water filter
		About 25 ft of 550 cord	

Southwest Desert Survival Gear Checklist —by Tony Nester

A quality survival kit is your life insurance policy during a backcountry emergency. Your kit should take care of the "Big 5" survival priorities of shelter, water, fire, signaling, and first-aid. I carry my own preselected items rather than using commercial kits. Here's a breakdown of my gear:

Knife—A knife is a critical survival tool, and I carry two blades: a Mora knife, which holds a good edge and can take a beating, plus a small, backup Cold Steel lockblade.

Fire Starters—Carry three, such as REI Stormproof matches, a spark magnesium rod, and a lighter. Fire = Life in the wilds under survival conditions, so carry these with you in your pockets and become proficient at making fire under any condition.

Tinder—Cotton balls smeared with Vaseline will practically allow you to make fire underwater. Make a half-dozen at home and place them in a film canister or an Altoid type container.

Emergency Blanket—These have grommets and will enable you to rig up a quickie shelter. An Army poncho works well too. Avoid the worthless Mylar blankets, as these will shred apart like tinsel in no time.

Water—Carry two to six quarts or more, depending on the time of year and length of hike. I like Nalgene bottles, which hold up to punishment on the trail. I also carry five gallons per person per day minimum in my truck.

Electrolyte replacement powder or capsules.

Water Purification Tablets- I use either Potable Aqua iodine or chlorine dioxide. Make sure to give it a taste test beforehand and follow the manufacturer's directions.

First Aid—A quality first-aid kit by either Adventure Medical Kits or Atwater Carey is essential. Add an ACE Wrap, Fastmelt Benadryl for insect bites, Imodium, and ibuprofen, and you will be ahead of the game if injury befalls you.

Signal Mirror—We have tested out glass signal mirrors in the desert and gotten reflections from twenty-six miles away. They are far safer, to you and the environment, than a risky signal fire.

Duct Tape—Imagine what our ancestors would have accomplished with this multiuse item. It's good for everything from patching up a torn pack to wrapping a blister to fixing a damaged boot sole. Wrap a few passes around your water bottle for quick access.

Flashlight—Get a quality LED headlamp for hands-free work when rigging up an emergency shelter in the dark or alerting searchers to your location.

Sunglasses/Sunscreen/Shemagh/Gloves/Brimmed hat—Enough said!

Jerky/Snacks/Chow—Yeah, you can survive without snacks, but it sure isn't fun.

Lastly, don't forget about that survival tool between your ears, and remember to leave a travel plan with someone back home so searchers know exactly where to look. That plan will be your safety net in the event that you run into Murphy's Law on the trail.

Desert Vehicle Recommended Equipment List

Roll of electrical tape	Emergency fuel
Flashlights (two) with extra batteries	Flares—at least six
Headlamp	Strips of carpet
Shovel	Extra fan belt
Spare tire	12-foot jumper cables
Jack	Axe or good hatchet
Tool kit	Food—dehydrated (requires
Tire chains	water)
Tire pump	Cooking pots
Vehicle repair manual	Tarpaulin for shade
Block and tackle	Blankets
50' of 5/8 inch rope	Poncho or sheets of plastic
Trekking shoes	

Moving in the Desert

When you're in the desert, learn to become nocturnal. Once the sun goes down, begin your activities.

Your first priority is to construct a shelter. Consider the type of shelter design you want to use, look the area over for possible sources of water, and determine how much water you have on hand. Do all of this from the shade of your temporary shelter.

Use your space blanket, casualty blanket, Tyvek, or poncho and some 550 cord to make a simple lean-to type of shelter. A casualty blanket, which is basically a quality NASA-designed space blanket, works especially well as shelter construction or for sleeping. To build a shelter, simply secure one end

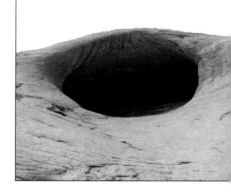

≈ **Desert cave**

≈ **Desert shelter** ≈

of the material to the ground using stakes or heavy stones, and angle the other end of the material upward in a manner that protects you from the sun.

Make the angled end no higher than four feet off the ground and secure it to bushes, stakes, or rocks. Then, place five or six pieces of light brush on the material and cover the entire shelter with material once more. Use Tyvek or a casualty blanket for the first layer of material and a space blanket as the top layer with the fluorescent orange side up to act as an emergency signal.

This type of "sandwich" shelter forms a dead air space between you and the sun. The insulation keeps the shelter cooler than a single-layered shelter. Construct this shelter in the cool of the evening and not during the heat of the day.

Vehicles

Traveling through the desert can be done successfully if a few simple precautions are taken. Be sure the vehicle(s) are in good working order, that fuel tanks and radiators are filled, that the batteries are well charged, and that all engine belts are in satisfactory condition. Also take a moment to inspect the tires and make sure they are inflated properly, and do not have abnormal wear spots.

Do not attempt to negotiate washes without first checking the footing and clearances, as high centers can rupture the oil pan.

If you get stuck, do not spin the wheels in an attempt to gain motion. Instead, apply power very slowly to prevent the wheels from digging in. When driving in sand, traction can be increased by partially deflating tires. Start, stop, and turn gradually; sudden motions will cause the wheels to dig in. Be sure to carry all tool and equipment requirements. (See the recommended desert survival items listed in this chapter.)

Roadway signs indicate civilization. So, if you find a road and it is safe to do so, stay on it.

If tactically safe, stay near your vehicle if it breaks down. Raise the hood and trunk lid to indicate that "help is needed." If you leave the vehicle, leave a note for rescuers with the time you left and the direction taken.

When not on the move, use any available shade or erect some shade from tarps or other material to protect yourself from the direct rays of the sun.

Know where you are at all times. When planning on entering unfamiliar country, always thoroughly study a map beforehand. Take special note of terrain features, the road structure, the direction to the nearest habitation, the location of water, and so on.

Upon arrival at your destination, look for landmarks and orient yourself with the prominent ones. As you move through the country, check your back trails often. Terrain always looks different when you are coming at it from the opposite direction.

Trekking

There are special rules and techniques for trekking in the desert. By moving slowly and resting ten minutes every hour or so, someone who is not injured and is in good physical condition can cover fifteen to twenty miles per day—less after becoming fatigued or suffering from lack of food or water. In the hot desert, it's best to travel in early morning or late afternoon into the evening, in whatever shade is available.

During your map study, select the easiest and safest route. Go around obstacles, not over them. Instead of going up or down steep slopes, zigzag to prevent undue exertion. Go around gullies and canyons instead of through them. When trekking with teammates, adjust the rate to the slowest member, who is often the one carrying the most gear or the injured.

On breaks, sit down in the shade and prop your feet up (to help reduce painful swelling), remove your shoes, and either change your socks (only if you're in the shade) or straighten out the ones you are wearing to reduce the chance of developing blisters.

If you get lost, sit down, survey the area, and take stock of the situation. Do not sit or lie on the ground because, in sunlight, the ground usually is thirty degrees F hotter than the air. Try to remember the last time when you knew exactly where you were. Decide on a course of action. It may be best to stay where you are and let your teammates or support assets come to you. This is especially true if you have access to water, fuel, or, in winter, some means of shelter.

If you feel you can retrace your course, do so. If safe to do so, mark your spot or leave a note before moving on. Look for your tracks; you may be able to backtrack and find the way to familiar ground. Be leery about taking shortcuts, as they may cause more confusion. If possible, go to a high point and look for distinguishable landmarks.

To avoid poisonous creatures, place your hands and feet only where you can see.

Footwear for Desert Hiking—by Tony Nester

Which footwear you choose depends on the desert. Three of our four deserts in North America are blanketed with cacti and other flora that can impale the careless. When hiking in the Sonoran, Mojave, or Chihuahuan deserts, I only wear lightweight leather boots. These allow me to avoid the spines of cacti, agave, and other succulents that can pierce. Sandals are out, except for walking around base camp at the end of the day.

The desert can be an unforgiving landscape if you're unprepared with the proper footwear. Geronimo and his Apache warriors had to mend their buckskin moccasin soles every three days or 100 miles of travel!

I have worn the Original S.W.A.T. brand boots on all of my desert treks and they have proven extremely reliable and affordable. I have also had good luck with the Hi-Tec brand.

If I'm going hiking in the Great Basin Desert amidst sagebrush then I would consider wearing lightweight hiking shoes. I currently have a pair of Merrells that have held up well.

With either boots or shoes, leather will hold up better than Cordura-type materials. Just remember to avoid black! I also carry a small comb in my pack, which can be handy for lifting out cactus and cholla spines in the boots rather than using my fingers, especially necessary if you hike with your dog.

Signals

One of the most effective signals is a fire—either a smoky one during daytime or a bright one at night. The signal mirror is also an excellent device for attracting attention, particularly aircraft. On a clear day, ground signals can be visible for up to ten miles; signals to aircraft can be seen at even greater distances.

If you decide to stay where you are and wait for rescue, it's a good idea to establish some type of ground-to-air signal, such as a large "X," "SOS," or the word "HELP." Use any available material to make this display (rocks, brush, clothing). It can even be scraped into the ground. The important thing is to change the terrain and attract attention to your location.

Signal sounds tend to be the least effective. Three sounds universally signifies "distress." A "thunder" whistle is recommended as an easy way to make loud noise. If you have a firearm, and it is tactically safe to do so, shoot once, wait ten seconds, then fire twice more, about five seconds apart. The first sound will hopefully attract attention, and the second and third will give direction.

≫ **Desert terrain**

Heat Disorders

Sunburn

Redness and pain. In severe cases—swelling of skin, blisters, fever, headaches. Use ointments on mild sunburns. If blisters appear, do not break them. If breakage occurs, apply dry sterile dressing.

Heat Cramps

These are painful spasms that usually occur in the muscles of your legs and abdomen. Sometimes they're accompanied by heavy sweating. Apply firm pressure on cramping muscles, or gently massage to relieve the spasms. Take sips of salt water (one teaspoon per glass) every fifteen minutes for one hour; better yet, take electrolytes.

Heat Exhaustion

Characterized by heavy sweating, weakness, and dizziness. The skin will feel cold, pale, and clammy. Pulse is usually steady and temperature normal. Sometimes accompanied by fainting and vomiting. Get the victim out of the sun, lie victim down, loosen clothing, apply cool wet cloths, and fan. Administer sips of salt water (one teaspoon per glass) every fifteen minutes for one hour or take electrolytes. If vomiting, do not give fluids or anything by mouth.

Heat Stroke

Symptoms include a body temperature of 106 degrees F or higher; hot, red, dry skin; rapid and strong pulse; and possible unconsciousness. Heat stroke should be considered a severe medical emergency. Seek medical attention ASAP, as delay can be fatal. Move victim into a cooler environment, reduce body temperature with iced bath or sponging.

⌃ **Cactus**

Water

Do not ration water. Rationing water at high temperatures can be an invitation to disaster, because small amounts will not prevent dehydration. In the hot desert, a person needs about a gallon of water a day. Loss of efficiency and collapse always follows severe dehydration. It is the water in your body that maintains life—not the water in your water containers.

If you drink more water than you actually need, it will pass in the form of urine. When you urinate, check the color and amount. Dark-colored urine indicates you need to increase your water intake. Many survival experts recommend drinking at least one quart of water for every two lost. But less fluid will not result in less sweat. In extreme heat, you may not even feel yourself perspire because the sweat evaporates so quickly.

Ration Sweat—Not Water

Keep your clothing on, including your shirt and hat, because it will help slow sweat evaporation and prolong cooling. It also keeps out the hot desert air and reflects the heat of the sun.

When day movement is necessary, travel slowly and steadily. Keep your mouth closed and breathe through your nose to reduce water loss and drying of mucous membranes. Avoid conversation for the same reasons.

Alcohol in any form is to be avoided, as it will accelerate dehydration. Food intake should be kept to a minimum if sufficient water is not available.

Dehydration

Body temperature in a healthy person can be raised to the danger point by either absorbing heat or generating it. Heat can be absorbed from the ground, by reflection, or direct contact. Work or exercise will also increase body heat. An increase in body temperature of six to eight degrees above normal (98.6F) for any extended period can cause death.

The body gets rid of excess heat and attempts to keep the temperature normal by sweating, but sweating causes the body to lose water and dehydration results. This water must be replaced.

Drink cool or warm water as fast as you want, but cold water may cause distress and cramps.

It's important to recognize the initial symptoms of dehydration. These include thirst and discomfort, slow motion, no appetite, and, later, nausea, drowsiness, and high temperatures. If dehydration reaches six to ten percent, symptoms may include dizziness, headaches, dry mouth, difficulty in breathing, tingling in arms and legs, bluish color, indistinct speech, and, finally, an inability to walk.

Thirst is not an accurate indicator of the amount of water that your body needs. If you drink only enough to satisfy your thirst, you can still dehydrate. Drink plenty of water, especially at meal times and during the cooler early morning hours. A pebble or small coin placed in the mouth will help to alleviate the sensation of thirst, but it is obviously not a substitute for water and will not aid in keeping your body temperature normal.

Water Procurement in the Desert

If you're near water, remain there and signal for rescuers. If water is not immediately available, look for it by following these leads:

- Look for desert trails—following them may lead to water or civilization, particularly if several such trails join and point toward a specific location.
- Look for flocks of birds—they sometimes circle over water holes. Listen for their chirping in the morning and evening, and you may be able to locate their watering spot. Pigeons or doves only exist near clear water.
- Look for animal tracks—they sometimes lead to water.
- Look for plants that only grow near water—cottonwoods, sycamores, willows, hackberry, salt cedar, cattails, and arrow weed. You may have to dig to find this water.
- Keep on the lookout for windmills and water tanks built by ranchers.
- An old Bedouin trick is to turn over half-buried stones in the desert just before sun up. Their coolness causes dew to form on their surface.
- Desert grass will also form dew in the predawn. It can be soaked up with a cloth and wrung out into a container.
- Where sand dunes meet the sea, digging above the high tide mark might reveal a thin layer of fresh water sitting atop a heavier layer of salt water.
- Flies and mosquitoes indicate a water source.
- Bees fly in a straight line to and from water up to 1,000 meters away.
- Water seepage in canyons, small pockets of water in sandstone rock formations, and digs at the base of rocks and mountains can produce water.

Methods of Purifying Water

Even contaminated water has its uses. It can be used to soak your clothing and reduce water loss from perspiration.

Dirty water should be filtered through several layers of cloth or allowed to settle. This does not purify the water, even though it may look clean. Purification to kill germs must be done by one of the following methods:

Boiling is the safest of available water disinfection methods. It kills *Giardia, Cryptosporidium,* bacteria, and viruses. At sea level, boiling water for one minute effectively eliminates these hazards, although vigorous boiling for two to five minutes is generally recommended for *Crypto sporidium.* At higher elevations, water boils at lower temperatures and longer boiling times should be observed (e.g., fifteen minutes at 10,000 feet).

Chemical Disinfection, usually with either chlorine or iodine, is another method of preventing infection from *Giardia* and most other micro-organisms. *Cryptosporidium* parasites are highly resistant to most chemical disinfectants, however, and can only be neutralized by boiling or filtration.

The table below lists the various disinfectants available and the recommended dosage per quart of water. The use of saturated iodine (made by dissolving iodine crystals in water) is not recommended because it does not kill all of the *Giardia* organisms in cold water. None of the below-listed disinfectants are considered to be effective against *Crypto sporidium*.

WATER DISINFECTION METHODS

Disinfectant	Quantity per Quart of Water	Waiting Time Before Drinking
Chlorine Tablets	5 Tablets	30 Minutes
Household Bleach	2 Drops	⋆ 30 Minutes
Iodine Tablets	2 Tablets	20 Minutes
2% Tincture of Iodine	10 Drops	20 Minutes
Saturated Iodine	Not Recommended	

When using tablets, the waiting time begins after the tablets are dissolved.
⋆Use 4 drops if water is cloudy or turbid

Using Bleach to Purify Water

Bleach is an oxidant, and it will react with and kill pretty much any microscopic cellular life (including viruses) that it comes in contact with. When it reacts, the bleach is actually consumed in the process.

Because killing microorganisms also consumes the bleach, the scent test tells you whether or not there's anything left to kill. If there's no chlorine odor, then all of the bleach was used up, meaning there could still be living organisms. If there is a chlorine odor after thirty minutes, it tells you that all of the bacteria and viruses are dead, and the bleach has done its job.

Most laundry bleaches have five and one-half percent Sodium hypochlorite, a suitable purification chemical for water. Bleach in a suitable container with an eyedropper dispenser makes a nice addition to any survival kit. Do not use powdered, scented, or other non-pure bleaches.

Prior to adding the bleach, remove all suspended material by filtration through a cotton cloth or by simply allowing sediment to settle to the bottom.

Add eight drops of bleach per gallon of water (or two drops per quart). If the water has already been filtered, shake it up for even dispersal of the bleach, and wait fifteen minutes. If it has sediment at the bottom, don't shake it up. Instead, allow the treated water to stand for thirty minutes.

Properly treated water should have a very slight chlorine odor. If you can't smell chlorine, repeat the dosage and allow the water to stand another fifteen minutes.

For cloudy, green, or really foul water (i.e.: swamp water), start with sixteen drops of bleach per gallon of water (or four drops per quart). As detailed above, smell the water. If there's a faint odor of chlorine, the water is drinkable. If not, then repeat the treatment.

Treating Larger Quantities of Water

- A teaspoon of bleach will treat about 7 ½ gallons of clear water or four gallons of dirty water.
- A tablespoon of bleach will treat about twenty gallons of clear water or about ten gallons of dirty water.
- A quarter cup of bleach treats about ninety gallons of clear water or forty-five gallons of dirty water.

Warning—Water from Natural Sources

During the past several years, increasing numbers of people have been stricken with waterborne diseases due to drinking water straight from natural sources—streams, springs ponds, or lakes. Even though the water appears to be sparkling clean and pure, it may contain microorganisms which cause disease.

Two organisms found in many water sources are *Giardia lamblia* and *Cryptosporidium parvum*. These parasites have been found in many wild and domestic animals and can be present even in very remote areas with no sign of human life. These organisms are transferred between animals and humans by means of excreted fecal material.

Drinking water containing these parasites can cause *Giardiasis* or *Cryptosporidiosis*. Both are severe gastrointestinal disorders, which can result in diarrhea, headache, abdominal cramps, nausea, vomiting, and fever. People with degraded immune systems should be aware that a *Cryptosporidium* infection can be life-threatening.

To prevent infection from *Giardia* or *Cryptosporidium,* do not drink naturally occurring water before you disinfect it. Either drink the water you carry in or disinfect all water from natural sources before drinking.

The minimum water consumption rate is two gallons a day per person in the 110- to 120-degree summer temperatures. In such extreme heat, I'd say that survival time without water would be limited to around two days, maybe less, depending on the variables mentioned above.

When it comes to water sources, don't assume that the creek, spring, or water hole you noticed on the map is going to even exist, especially during a season of drought. Talk to the folks who are out on the land all the time—the locals in the area—and find out what the water conditions are really like in the backcountry.

On an extended desert trip, and certainly in a survival situation, it's important to know how to locate water.

Being able to read the nuances of the land is a skill of visual acuity. You're searching for subtle clues written across the terrain that may indicate water. This is a skill that comes with experience.

Places to Look for Water

- Shady areas at the base of cliffs
- Rock pockets and depressions
- Tree cavities and hollows
- Undercut banks in dry riverbeds
- Where insect life abounds
- Animal tracks and bird signs
- Where vegetation abounds: willows, palm trees, and cottonwood trees are water-loving trees found in proximity to either surface or subterranean water.

Remember, a hike to a suspected water source is going to cost you, in terms of your own precious sweat, so staying put might be a better option if rescuers are on their way.

Hyponatremia or Water Poisoning

When you're sucking up quarts of water each hour in the intense heat, you have to account for lost electrolytes. Most people are familiar with the dangers of heat exhaustion but not familiar with hyponatremia, which results when you are technically hydrated but essentially flushing the electrolytes from your system every time you urinate. This is something that afflicts many novice desert trekkers.

For every thirty minutes of activity in the intense heat, take a shade break, rehydrate, and get in some type of electrolyte replacement.

The Problem with Solar Stills—by Tony Nester

In my opinion, solar stills are not a good way to procure water in the desert.

On every desert survival course, we construct a solar still just to show the futility of this method. After digging a three-foot-deep hole that is three feet in diameter (and we all carry shovels in our packs), we line the pit with succulent, nontoxic plants (grasses, cacti, and so on) to boost the output, place a cup in the bottom, and then seal it up with an inverted six-foot by six-foot sheet of clear plastic. When possible, we construct the still in the damp soil of an arroyo or canyon floor that has recently seen rain. The next day, after the still has had twenty-four hours to work in the sun, we pull back the plastic cover and voila!—there's about a half quart of clear fluid! Wow! Then we remember back to the previous day when we burned off a gallon of sweat making the still.

The single best method I know for staying hydrated in the desert is to be prepared and carry water with you.

The question often arises, "If I run out of water, should I push on in search of more or should I stay put and go without water?" There will be a tremendous physiological cost if you tax your heat-stressed body further by searching for water, especially during the peak hours between 10 AM to 4 PM.

If you finish your water, it very well might make sense to hole up in the shade like a coyote, all the while staying clothed to reduce evaporative sweat loss, and traveling, if necessary, only during the cooler hours of the morning or evening—if at all. There have been survivors lost in the desert, without water, who have endured up to two days in triple-digit heat by doing the above. There have also been those who perished from heat stroke in three to four hours while hiking in search of water. You really have to weigh the situation at hand to determine if staying put or hiking on is required. In most cases you will want to stay put to extend your survival time.

(For more about water, see Chapter 8, page 131.)

⌃ **El Centro, Calif—Water is the most vital resource for surviving in the desert. Students in Desert Environmental Survival Training (DEST) use a water procurement method of evaporation to extract drinkable water from saltwater. The saltwater in the hole evaporates leaving the salt at the bottom, but with a plastic cover over the hole, the water has nowhere to go and drops back down into the jar as potable water.**

Desert Food

If you don't have enough water, you shouldn't be eating. When your body processes food into waste, fluids from your body are used. So, if you don't have enough water in you, you can actually speed up dehydration by eating. Most people can go for significant periods of time without eating. Water is your primary concern in the desert—not food.

Edible Plants—All cactus fruits are safe to eat, but other plants and legumes (bean-bearing plants) can be poisonous depending on the season and habitat. (There are more than 700 poisonous plants in the United States and Canada alone.) The best advice is that, unless you're an expert on plants in a particular area, don't consume them. Remember that you must have water, but you can go without food for days without harmful effects.

In a survival situation, where use of strange plants for food is necessary, follow these rules:

* Avoid plants with milky sap.
* Avoid all red beans.
* If possible, boil plants that are questionable.
* To test a cooked plant, hold a small quantity in your mouth for a few moments. If the taste is disagreeable (very bitter, nauseating, burning), don't eat it.

≈ **Prickly pear cactus**

≈ **Yucca**

Edible Plants in the Southwest—by Tony Nester

There are wild plants that are truly "palatable" (these were the good-tasting staples used by native cultures) and then there are the "edible" plants (these were survival foods used in lean times and that have a considerable gag factor).

In the southwestern United States, the Hopi Indians have identified 150 plants, of which roughly twenty-nine are suitable for eating, and another forty can serve some medicinal purpose.

When it comes to harvesting wild plants, be certain you know what you are picking and putting in your mouth.

The fishhook barrel (*Ferocactus wislizenii*) contains water, but I would hardly call the nasty fluid extracted from its innards "water." It's high in alkaloids and has a considerable gag factor, and there are four other barrel cacti that are toxic.

The few times I have tried choking down barrel cactus fluid, it made my stomach churn like a cement mixer and required a meditative concentration not to regurgitate.

Piñon Pine Nuts (*Pinus edulis*)—During a good fall season, the pinon pine can yield a significant amount of protein-rich nuts. Whether you roast them or eat them raw, you will want to remove the delicate shell to liberate the sweet meat inside. One study found that a single pound of pine nuts will yield 3,000 calories, so this is a very worthwhile food source.

Prickly Pear (*Opuntia spp.*)—The bright green, spineless pads are collected in the spring and sautéed, boiled, or steamed. I like to gather them when they are the size of a silver dollar, as larger ones become too stringy. The red fruits in late summer can be collected once the tiny stickers or glochids are removed by peeling the skin off or lightly charring the fruits for a few minutes in the hot coals of a campfire. These fruits are sweet but heavy with seeds. For preserving the fruits, cut them in half and dry in the sun over a few days to make cactus fruit jerky.

Yucca (*Yucca baccata*)—The large fruits from banana yuccas can be collected when ripe in late summer and roasted on coals for an hour. The modern-day forager can simply wrap them in foil to bake in a conventional oven at 400 degrees until cooked through like a potato. If it tastes like you're eating a bar of soap, then you collected the fruits too early in the season.

Pine Needle Tea—superhigh in vitamin C, this tea was used by the mountain men to treat scurvy. Simmer a handful of diced, green needles in a cup of water for twenty minutes. Do not boil as that will destroy the vitamin content. This tea can taste like pine sap.

❯ **Pine needles**

❯ **Cricket**

Insects—Generally you can eat bugs. Do not eat scorpions, centipedes, or brightly colored insects.

Spiders—Avoid eating any kind of spider.

Snakes—Snakes can provide a filling meal. The sheer amount of bones to sort through in a snake makes for a lengthy meal, however. Trying to kill a snake is a last-ditch survival effort and one that I don't recommend. Food is not a short-term survival priority and many have gone thirty to forty days without food in the desert.

It's non-venomous snakes that a survivor would want to ideally procure. This is best done with what I call the Grady Gaston Method. Grady was a member of a B-24 bomber in WWII that crashed in Australia after

❯ **Timber rattlesnake**

⩘ Brown recluse spider

⩘ Black widow spider

returning from a bombing run against the Japanese in New Guinea. He survived more than 130 days, most of it solo, while living on snake meat that he obtained by hurling large rocks on the creatures. By the way, he didn't have the means to make fire, so he consumed these raw! In the end, it was his sheer willpower, sense of optimism, and serpentine diet that enabled him to endure one of the most epic tales of survival in recent times.

Beware of rattlesnakes, as they can still bite you after they're dead due to a reflexive action of the nervous system. Lopping the head off, burying it, and then skinning and cleaning the snake are the recommended methods. Once skinned and cleaned, the meat can be boiled up in a stew with any other tasty tidbits or edible plants, or placed on sticks and cooked shish kebab style over the coals.

Venomous Creatures

Carefully inspect all clothing and bedding before use, especially items that have been on or near the ground during the night. Dampness attracts these creatures. During summer evenings, scorpions travel over the desert floor and up the branches of trees and bushes looking for food. Bedding on the ground will provide them a hiding place toward morning.

Spiders—Deserts have many kinds of spiders, but most of them aren't venomous. The two to watch out for are the brown recluse and black widow.

Brown recluses are light brown, about a quarter inch in length, and have a violin-shaped marking on the head and back. They are most active at night. If you're bitten by a recluse and don't get medical attention, it could be very serious. The bite causes severe tissue destruction that may take weeks to months heal. In extreme cases, the bite can be fatal.

The female black widow spider is the most poisonous spider in North America. It is easy to recognize) by the red hourglass shape on the underneath part of her abdomen. She has a shiny black body with various types of red markings on the top, depending on the species. There are about five species of black widow spiders in North America. They are usually found in the dark corners of sheds and outbuildings, under logs, and in rock piles. Will bite if provoked. The bite can be dangerous but is seldom fatal. Pain spreads throughout the body, accompanied by headache, dizziness, and nausea. Extremities become cramped, the abdomen becomes rigid, pupils dilate, and spasms may occur after several hours.

Lizards—There are two that are venomous—the Mexican beaded lizard and the Gila monster. The Mexican Beaded is only found in the deserts of Mexico and Guatemala. Gilas are found in the Sonoran Desert of the United States and northern Mexico.

The Mexican Beaded has white to yellow spots and stripes on round, raised scales and is about a foot long. Gilas can be as large as two feet and have round, raised scales. They're short, stout, and have a thick tail. While both are venomous, neither is considered fatal.

Treatment for Bites and Stings:

- Seek shade.
- Sit down and try not to move the affected limb.
- Wash the bite with soap and water.
- Elevate the affected area above your heart level.
- Do not lance the bite or attempt to suck out the venom.
- Remove any jewelry near the bite and loosen tight clothing.
- Tie a light constricting band around the affected body part about three inches above the point of contact. Keep it loose enough to fit a finger between the band and your skin (to prevent it from becoming a tourniquet).

Honeybees—There are more deaths annually in the U.S. from honeybee stings than from all other poisonous creatures combined. Honeybee stingers are barbed at the tip and will remain in the victim. The venom sacs are torn from the bee's body and remain attached to the stinger. Don't try to pull out

≽ **Gila monster**

≽ **Non-venomous lizard**

≈ Bee

≈ Scorpion

stingers, as pinching them injects additional venom. Scrape stingers out with a knife or other thin edge.

Africanized Honey Bees or "Killer Bees"—They resemble U.S. and common European honeybees, but differ in temperament. Africanized honey bees defend their colonies more vigorously and in greater numbers. They may respond with minimal or no provocation, but their venom is no more harmful than that of domestic honeybees. If you see a lot of bees flying in and out of a small opening, a nest is probably located inside. The best strategy is to leave them alone and do not disturb them. If attacked, run away as fast as you can. If far from shelter, try running through tall brush. This will confuse and slow them while you make your way out of the area. Do not flail or attempt to swat the bees. Bees target your head and eyes; therefore, try to cover your head as much as possible without slowing your progress. If someone else is stung by honeybees, help them out of the area as quickly as possible.

Rock or Bark Scorpion—These are small, very slim, and light straw colored. The stinger in the tip of the tail injects a minute amount of powerful venom. There will be pain at the site, numbness, restlessness, fever, fast pulse, and breathing difficulty. The sting can be fatal.

If bitten or stung:

- Sit down in the shade and relax.
- Wash the area with soap and water.
- Apply a cool compress to the bite or sting.
- Elevate the area above your heart level.
- Take an over-the-counter pain reliever.
- Tie a light constricting band around the affected body part about three inches above the point of contact. Use a bandanna, a shoestring, some gauze, or anything else that isn't too heavy. Again, it should be loose enough to get a finger between your skin and the wrap.

Rattlesnakes—In the desert, rattlesnakes are generally sandy colored with a broad arrow-shaped head, blunt tipped-up nose, and rattles on the tail. Look for them mostly where food, water, and protection are available—around abandoned structures, irrigation ditches, water holes, brush, and rock piles. They don't always warn by rattling, nor do they always strike if someone is close. Usually they are not aggressive and will not "chase" people. If you are bitten, the strike results in immediate pain accompanied by swelling. The venom primarily causes local and internal tissue destruction and nerve damage. If traveling in areas where rattlers may be found, wear protective footgear and watch where you put your hands and feet. The general rule of thumb is if you hear one, stop and try to locate it, then move slowly away from the sound and leave it alone. Most strikes occur when people attempt to catch, kill, or play with the snake.

Coral Snakes—Rarely more than twenty inches long with a small blunt, black head, and tapering tail. A very attractive snake with wide red and black bands, separated by narrower yellow or white bands that completely encircle the snake. There are many nonpoisonous species that resemble the coral snake. Remember "if red (band) touches yellow—kill a fellow; if red touches black- venom lack." They're sometimes seen in the day in spring (March, April, and May) and are nocturnal during the summer. They live under objects, in burrows, and are shy and timid. Corals must chew rather than strike to introduce venom, but due to the very small mouth are unable to bite any but the smallest extremities. They attack only under severe provocation or accidental contact. Their venom affects the nervous system, causing failure of the heart and respiratory muscles.

Treatment of Poisonous Snakebites

- Elevate the limb that has been bitten and immobilize it.
- Apply a light constricting band above the bite location (be able to insert two fingers under band). Don't release the band unless it becomes too tight from swelling.
- Identify the snake and relate this information to the medical responders.

Flash Floods—by Tony Nester

≈ **Coral snake**

The key to avoiding an intimate encounter with a flash flood is to plan ahead so you don't get caught in one in the first place. Most research indicates that eighty percent of flash floods happen between noon and 8 PM during the monsoon seasons. Try to avoid narrow canyons during this time altogether or, if you must hike in them, check the weather first and then make movement during the early morning hours before the afternoon thermals reel in the moisture and thunderheads.

Flash floods, and the thunderstorms that cause them, are deceiving. You may be trekking in a canyon with blue skies above while there's a storm cell ten miles up canyon dumping its energy. Now you have not only a huge wall (five-plus feet or higher) of water headed your way, but also silt, rocks, logs, and other debris that has built in the canyon since the last rainfall, which could have been years or even decades ago. I have seen van-sized boulders tumbling down canyons like marbles during flash floods. Flash floods are the number one weather-related killer of people in the desert the world over.

A Short Survival Exercise with SERE/CSAR—by Tony Nester

It was nearing 106 degrees F at noon when we headed out for an E&E exercise in the Arizona desert. This was the final phase of a desert training course that I was conducting for a group of SERE and CSAR personnel.

A route five miles ahead was selected and the UTM coordinates were locked in on everyone's GPS units. We would split into two teams and rendezvous at a lone rock pinnacle at the five-mile mark, while each team attempted to reduce signs of their passage with evasion shelters previously taught.

The men were carrying their normal packs and combat vests, which together came to around sixty-five pounds with water (four liters each). Food rations consisted of three survival bars a day (1,200 calories total a day).

As we began, we were all silently grateful the sky was growing overcast and cooling things down as an impending storm to the south was brewing.

After resting, rehydrating, and checking everyone for heat-related injuries, we pushed on another three miles. The storm that was at our backs on the trek in was now in full force behind us and brought in 40 MPH winds that created a sandstorm. Out came the goggles and shemaghs. With little respite from the wind and with the swirling sand obscuring distant landmarks, we navigated in short increments from one boulder outcropping to another, resting every mile or so. During rest breaks, out came the rations and the sodium-replacement drinks (premixed into the water earlier). Lightning was in the distance as we pushed on, and each man was told to spread out thirty feet from the next person in observance of lightning safety.

By 4 PM, our destination was reached and we holed up in a large boulder-strewn region out of the wind. Water was procured from nearby *tinajas* or sandstone depressions in a nearby arroyo. A test shelter in the ground was dug and covered with a Mylar blanket. Everyone slept that night near or under a rock overhang in the boulders. The storm cleared out near sundown and the wind calmed. Night time temps were in the mid-forties. Any colder and a fire would have been necessary—a difficult prospect in so barren a region.

One student awoke to find a large rattlesnake on the other side of his pack. His startled demeanor scared the snake away. It was time to head back, debrief on the survival gear tested, and wash the sand out of our ears. Not a cloud in the sky today—it's going to be a furnace again!

Lessons Learned

Combat vests were shed after two miles and stowed in the rucksacks. Detailed water consumption records were kept and each man went through approximately six liters during daylight hours of activity. The allotted food rations were sufficient for the two-day time frame, but everyone noticed slight gastrointestinal stress after the first day of consuming only survival rations and water.

Footwear was a definite weak link, with each man suffering considerable blisters and bruised soles from inadequate (issued) boots. Desert footwear for this group needs to be improved. Danner Desert boots or Original S.W.A.T. boots, used by our instructor cadre with good results over the years, were recommended.

The Mylar-type blankets were also inadequate and shredded after only a few hours' use. An Emergency Blanket, poncho liner, or camo poncho would be far more useful and durable enough to withstand the repeated rigors of survival on the move while not adding significant weight.

Other questions to consider: How will my current issued gear hold up if I have to escape and evade over rough terrain for a few days or more? Have you tested, under actual field conditions, the limitations of the survival items that you are issued? How far can you run/hike in the sand with a full ruc before needing to shed gear, and what essential gear can you absolutely NOT live without for survival in the desert? Do you know some of the basic techniques for covering your tracks and gaining distance from pursuers? Do you know how to make a smokeless, concealed fire? Have you lived exclusively on your issued survival rations in the field before to test out their usefulness in maintaining? How does your present footwear hold up over rough, rocky terrain?

Survival in the Outback

In April 2006, Mark Clifford, a farm manager on a remote property in Australia's Northern Territory, thought he was seeing a walking skeleton coming toward him. It turned out to be a 35-year-old man named Ricky Megee, who had been lost in the outback for an incredible ten weeks.

Apparently drugged and left for dead by a hitchhiker he'd picked up, Megee, who had nothing but the clothes on his back, survived by staying close to a dam and eating leeches, grasshoppers, and frogs.

While police and the public had doubts about the story because of Megee's previous minor drug convictions, there's no question that he had been lost in the outback, for whatever reason, and was damn lucky to have survived.

Wandering for weeks, he eventually found a water hole, where he spent the remainder of his time. Baked in the day and frozen at night, he survived at first by eating frogs.

"Meat of any description was the substance I craved," Megee explained later. "I was prepared to do whatever was required. But without that opportunity of a carnivorous feast always available to me, I sure ate a lot of vegetation. Edible plants didn't stand a chance with me around. "Crickets were the first really crunchy things that I tried, but definitely not the last." According to Megee, they made a nice change from the softness of the mushy vegetation he'd become accustomed to. He'd pull off their heads and chew the rest down as fast as he could so he didn't have to dwell on what he was actually eating.

"Grasshoppers were pretty crunchy as well," Ricky said. "I didn't really appreciate the sensation of the legs and wings tickling the insides of my cheeks—they were too spindly for my liking. To counter that, I pulled off their limbs and just ate the body, which made them more palatable."

Megee decided to try any type of food that didn't look or smell like it could kill him. Leeches surprised him, because he found them easier to eat than he had expected. "For anyone who hasn't tried leeches, they are some of the sweetest tucker you're ever likely to find when lost in the outback," he explained. "I'd heartily recommend them, as long as you learn how to chew fast."

He discovered that his mind-set was important. "If I complained about anything," Ricky said after this ordeal was over, "suddenly the frogs and leeches wouldn't want to know about me. But if I praised God for giving me nothing, then all the luck seemed to come my way the next morning."

Food, water, and shelter were his basic requirements for survival. But as the days wore on and Megee grew weaker, he had to look for eating alternatives that didn't require expending so much valuable energy.

He even tried to eat one of the cockroaches that had invaded the little mud shelter he'd constructed. "Just bringing that putrid, disgusting thing near my mouth created a smell strong enough to make me want to spew," Ricky explained later. "But there was still the remote possibility they tasted like peaches—I had to go through with it."Ignoring the stench, he shoved the cockroach headfirst into his mouth. But even a starving man has his limits.

"I'm not sure if it was the stomach-churning taste or the smell that got me in the end," Megee said. "However, the result was putrid enough to have me hurling uncontrollably out of the end of my shelter within two seconds. I didn't even manage to chew on him in the end—he was spat out before he sucked his last breath."

Despite eventually losing half of his body weight, Ricky Megee survived an incredible ten weeks. He did so by instinctively solving the basic requirements of water, food, and shelter—and adopting a survival mind-set that pulled him through.

⌃ **Desert sunset**

6

SURVIVAL AT SEA

"Success is not final, failure is not fatal. It is the courage to continue that counts."

—*Winston Churchill*

Water covers approximately seventy-five percent of the earth's surface. Seventy percent of that is made up of seas and oceans. Assuming you will cross these vast expanses of water in your lifetime, there's always a chance that a crippled boat or aircraft will make you lost at sea.

Survival at sea is especially challenging and will depend on the rations and equipment you have available, your ingenuity, and your will to survive. You can expect to face waves, high winds, and possibly extremes of heat and cold.

Learn How to Use Available Survival Equipment

Whether traveling by boat or plane, take time to familiarize yourself with the survival equipment on board. Find out where it's stowed and what it contains. Ask yourself: How many life preservers, lifeboats, and rafts are there? Where are they located? Are they stocked with food and medical equipment? Familiarize yourself with exits and escape routes.

Aircraft

If you're in an aircraft that goes down at sea, get clear and upwind of the aircraft as soon as possible. Stay in the vicinity of the aircraft until it sinks, but clear of any fuel-covered water in case it catches fire.

Look for other survivors. If they're in the water and you're in a lifeboat or raft, throw them a life preserver attached to a line, or send a rescuer from the raft with a line secured to a flotation device that will support the rescuer's weight (and help that person conserve energy). It's very important that the rescuer always wear a life preserver.

Be careful how you approach a panic-stricken person. Try to approach any survivor who needs to be rescued from behind. If possible, grab the back strap of the survivor's life preserver and pull that person to the closest available lifeboat or raft by swimming sidestroke.

If you're alone in the water and no rafts are available, find a large piece of floating debris and try clinging to it or even using it as a raft.

Understand that floating on your back expends the least amount of energy. Spread your arms and legs, arch your back, and lie down in the water. Your body's natural buoyancy will keep the top of your head above the water. If you relax and breathe evenly in and out, you can keep your face above water and even sleep in this position for short periods of time.

If you're unable to float on your back or the sea is too rough, float facedown in the water.

Get to a raft or lifeboat as quickly as possible.

Once you're in a raft or lifeboat:

1. Give self aid and first-aid if necessary to others onboard.

2. Take seasickness pills, if available, by placing them under your tongue and letting them dissolve. Remember that vomiting caused by seasickness increases the danger of dehydration.

3. Salvage all floating equipment, including rations, containers, clothing, seat cushions, parachutes, or anything else that can be useful. But make sure that these items contain no sharp edges that can damage or puncture your lifeboat or raft.

4. If you're in the vicinity of other rafts or lifeboats, lash them together so that they're approximately 7 ½ meters apart. That makes it easier for an aircrew to spot you.

5. Use all electronic and visual signaling devices to make contact with rescuers.

6. Check to see if there's an emergency radio or other signaling devices onboard. If so, activate it immediately. If you're in enemy territory, use these devices only when you think friendly aircraft are nearby.

7. Wipe away all fuel that might have spilled on the raft because petroleum will weaken the raft and erode its glued joints.

8. Check the inflation of your raft regularly. Chambers should be full, but not tight. Remember that air expands in heat. So on hot days, air might have to be released, and the chambers inflated in cooler weather.

9. Try to stay close to the crash site so that you're easier to locate by rescuers. You can do this by throwing out the sea anchor or by improving a drag with a bailing bucket or a roll of clothing. When you deploy the sea anchor, make sure that it's open because a closed anchor will form a pocket that will help propel the raft with the current.

10. Wrap the anchor rope with cloth so it doesn't chaff the raft.

11. Keep your raft as dry as possible, with everyone seated, and the heaviest passenger in the center.

12. Waterproof items that might be affected by saltwater— i.e., watches, compasses, matches, and lighters.

13. Ration food and water.

⩽ **Gingerroot settles the stomach and can help with sea sickness.**

14. Together with the other survivors, take stock of your situation and supplies, and plan for what it's going to take to survive.

15. Assign duties to each person—i.e., water collector, lookout, radio operator, bailer.

16. If you're in unfriendly waters, wait until nightfall before paddling or hoisting a sail. Be sure aircraft are friendly before trying to signal and get their attention.

17. If you're in a cold climate, rig a windbreak, spray shield, and canopy. Stay dry and insulate your body as much as possible, including protecting yourself from the cold bottom of the raft. Huddle with others to stay warm. Remember that hypothermia occurs rapidly when you're immersed in cold water because of the decreased insulating value of wet clothing.

18. If you're in a hot climate, rig a sunshade or canopy, leaving room for ventilation. Cover your exposed skin and protect it with sunscreen if available.

Boat

If you have a supply of food, fishing equipment, and fresh water, you can survive for a long time in a boat. If you don't have enough fresh water, make a rainwater collection system with a tarp or raincoat that runs into a container. Or simply place collection containers on the deck during rainy weather. Drink at least a liter of fresh water a day, fish a little, and try to relax.

≫ **Life raft**

Life Raft

Life rafts are a lot like boats but have a greater chance of sinking due to punctures, leaks, rips, or defects. Modern life rafts are durable and come well-equipped for emergencies. They can range from a one-man raft to larger 25-man rafts. They're usually equipped with some combination of the following:

* Covered deck
* Paddles
* Insulated flooring
* Bailing buckets
* Ladders
* Flares
* Water collection pouches

- Signaling mirrors
- Reflective tape
- Fishing kits

In the Water

If you're in the water without a boat or raft, you have your work cut out for you. Wear an inflatable safety vest, if available. It will keep you floating. If you're in cold water, pull your knees to your chest, which will help your body retain heat and resist hypothermia.

If you can't swim well but need to cross a large body of water, use your pants as a flotation device. Simply remove your pants, tie off the legs, then allow them to fill them with air (a skill often taught in U.S. military boot camp—all services). Raise the pants over your head in the water and they'll act like a life jacket.

Drinking Water

Drinking water is vital to your survival. With it alone, you can survive for ten days or more. When you consume it, wet your lips, tongue, and throat before swallowing.

Protect your freshwater supplies from saltwater contamination and use it efficiently. Calculate daily water rations by measuring the amount of fresh water you have, the output of solar stills and any desalting kit, and the number and physical condition of the people on your lifeboat or raft.

If you run out of fresh water, don't eat. Do not drink seawater or urine!

To reduce water loss through perspiration, soak your clothes in seawater and wring them out before putting them on again. But do this only when necessary because you can develop saltwater boils and rashes from wet clothes.

Keep a clean tarpaulin ready to catch water from showers. A small amount of seawater mixed with the rainwater you catch is not a cause for concern. The water will still be safe to drink and won't cause a physical reaction.

At night, hang the tarpaulin like a sunshade and turn up the edges to collect dew. Dew can also be collected from the sides of the lifeboat or raft using a sponge.

If solar stills are available, set up the stills immediately.

If you have desalting kits in addition to solar stills, save the desalting kits for overcast periods when you can't use solar stills, or catch rainwater.

In arctic areas, old sea ice, which is bluish, has rounded corners and splinters easily, is nearly salt free, and can be used as a source of water. Avoid new ice, which is gray, milky looking, and hard. Water from icebergs is fresh, but should only be used in case of an emergency. Icebergs are extremely dangerous.

If you run out of fresh water, you can drink the aqueous fluids found along the spine and in the eyes of large fish. Cut the fish in half to get to the fluid along the spine, and suck the fluid of the eye. Avoid any other fish fluids, as these are rich in fat and protein and will use up more water during digestion than they will supply.

Sleep and rest are the best ways to endure periods with little or no food and water.

Food

When at sea, fish will be your main source of food. With a few exceptions, fish caught out of sight of land are safe to eat. Fish nearer to shore are more likely to be poisonous. Some fish that are normally edible—like red snapper and barracuda—are poisonous when caught in atolls and reefs.

When fishing with a line, avoid handling it with bare hands or tying the line to your lifeboat or raft. Use gloves, if available, or a cloth to handle fish.

Don't eat fish that have pale, shiny gills; sunken eyes; flabby skin; or an unpleasant odor. Good fish smell like saltwater.

Cut fish that you don't eat immediately into long, thin strips and hang them to dry. Dried fish will remain edible for several days. Fish that hasn't been cleaned or dried will spoil in half a day. In warm areas, gut and clean fish immediately after catching them.

The heart, blood, intestinal wall, and liver of most fish are edible. Also edible are the partly digested smaller fish that you may find in the stomachs of large fish. Sea turtles, eels, and sea snakes are edible. But be careful when handling sea snakes because their bites are poisonous.

All shark meat is edible—either raw, dried, or cooked—except for the Greenland shark, which contains high levels of vitamin A. Shark meat spoils rapidly, so bleed it immediately and soak it in several changes of water.

Fishing

Fishing line can be made from shoelaces, parachute suspension line, or pieces of thread that have begun to unravel from tarpaulin or canvas. Simply tie them together until you have a usable fishing line.

If you have a grapple, or can improvise one out of available materials, use it to hook seaweed, which can contain crabs, shrimp, and even small fish.

Seaweed itself is edible but contains salt. Only eat it if you have sufficient drinking water.

Use small fish remains and the guts from birds as bait. A net improvised from cloth can be used to catch small fish.

Be careful not to puncture your lifeboat or raft with hooks or sharp instruments. And be careful not to capsize by trying to catch large fish.

Don't fish when you see large sharks. If you see a large school of fish, try to move closer to them.

At night, light attracts fish. So if you have a light, try fishing at night.

During the day, fish are attracted to shady areas. You might find them under your lifeboat, raft, or floating seaweed.

A spear made out of a knife tied to an oar blade can be used to spear large fish. Tie the knife tightly so that you don't lose it, and get large fish into your craft quickly so that they don't slip off the blade.

⌃ **Shoelaces could be used in place of fishing line.**

Birds

All birds are edible. Try towing a bright piece of metal behind you to attract birds. It's possible to catch birds with your hands or with a noose. Bait the noose in the center with a piece of fish if available, and wait for the bird's feet to enter before you pull it tight.

All parts of a bird are usable. The feathers can provide insulation, and the entrails and feet can be used as bait.

Health Issues and Hazards

Seasickness

The motion of your lifeboat or raft can result in nausea and vomiting, which in turn can lead to dehydration, exhaustion, and loss of will.

Treat seasickness by not eating food until the nausea is gone. Lie down and rest and take seasickness pills if available.

Some survivors have reported that erecting a canopy or using the horizon as a focal point has helped them overcome seasickness.

Saltwater Sores

These are caused by a break in skin that has been exposed to saltwater for an extended period of time. If scabs and pus form, do not open or drain. Flush the sore with fresh water and allow to dry. Apply an antiseptic if you have one.

⩔ Shark ⩔

Sunburn

Try to stay in the shade and keep your head and skin covered. Especially vulnerable are backs of ears, the skin under your chin, and eyelids. Remember that sunlight reflects off the water, so you get sunburn not only from above but from below.

If glare from the water causes your eyes to become bloodshot and inflamed, bandage them lightly.

Constipation

Do not take a laxative, as this will cause further dehydration. Exercise as much as possible and drink water.

Sharks

Sharks will present the greatest danger. Whales, stingrays, and porpoises might appear threatening, but they pose little danger in the open sea.

Consider any shark more than one meter long to the dangerous. But keep in mind that out of hundreds of shark species, only about twenty of them are known to attack man. Those that are dangerous include the great white shark, hammerhead, mako, tiger, gray, lemon, sand, nurse, bull, and oceanic whitetip shark.

Sharks found in tropical and subtropical waters are more likely to be aggressive. They possess an acute sense of smell and become excited by the smell of blood. They're also very sensitive to vibrations in the water.

Sharks feed all day and night, but most attacks on humans have occurred during daylight and especially the late afternoon.

When you're in the water, stay with other swimmers, watch for sharks, and avoid urinating, defecating, and vomiting in the water.

If you're in the water and feel that a shark attack is imminent, splash and yell to try to keep the shark at bay. If you're attacked, kick and strike the shark on the gills or eyes.

If you're in a raft or lifeboat and see sharks, stop fishing, don't throw garbage overboard, keep quiet and stop moving around.

If a shark attacks your raft, hit it with anything you have except your hands.

Hypothermia

Body thermal conductivity in water is twenty-six times faster than when exposed to air. If you have a life raft, board as soon as possible.

Hypothermia Chart		
WATER TEMPERATURE °(F)	**EXHAUSTION OR UNCONSCIOUSNESS**	**EXPECTED TIME OF SURVIVAL**
32.5	Under 15 Minutes	Under 15–45 Minutes
32.5–40	15–30 Minutes	30–90 Minutes
40–50	30–60 Minutes	1–3 Hours
50–60	1–2 Hours	1–6 Hours
60–70	2–7 Hours	2–40 Hours
70–80	3–12 Hours	3 Hours—Indefinitely
80+	Indefinitely	Indefinitely

Chart and facts courtesy of Winslow Life Raft

The use of an immersion suit or other buoyant thermal protective device will greatly enhance survival time. (This chart is for general reference only.)

Detecting Land

Deep water is dark green or dark blue. Shallow water—which might mean that land is nearby—is usually lighter in color.

In the tropics, reflected sunlight off shallow lagoons or coral reefs often gives a greenish tint to the sky. In the arctic, light-colored reflections on clouds often indicate ice fields or snow-covered land.

If you see a fixed cumulous cloud in a clear sky or in a sky where other clouds are moving, it's often hovering over or downwind of an island.

Birds are another indicator that land is nearby. The direction that flocks of birds fly at sunrise or dusk may indicate the location of land.

Nighttime fog, mist, or rain may carry with it the smells and sounds of land.

Approaching Land

Rafting ashore in strong surf can be dangerous. If you have a choice, avoid a nighttime beach landing.

Try to land on the leeward side of an island or on a point of land that juts into the water. Look for gaps in the surf line and steer toward them. Avoid coral reefs, rocky areas, rip currents, and strong tidal currents.

If you're approaching the shore through surf, take down the mast (if you have one), put on clothes and shoes, and inflate your life vest. Trail the sea anchor using as much line as you have, and use the oars or paddles to adjust the anchor to keep its line taut. This will kept the raft pointed toward shore and prevent the current from pushing the stern forward and capsizing you.

Steer to the sea side of large waves, which will help ride you in. If you're facing strong winds and heavy surf, you have to move the craft rapidly through the oncoming crest to avoid being turned broadside or thrown end over end. Try to avoid meeting large waves at the moment they break. If the surf is medium with no wind or an offshore wind, keep the craft from passing over the waves too rapidly so it doesn't drop suddenly after topping the crest.

When you near the beach, ride in the crest of a wave and paddle in as far as you can. Don't get out of the craft until it has grounded. Then jump out quickly and beach it.

In the unlikely event you feel conditions make it impossible to make it ashore via craft, jump out of the boat and sidestroke or breaststroke ashore wearing your shoes and clothing. In moderate surf, you can ride in on the back of a small wave by swimming forward with it. If the surf is strong, swim toward shore in the trough between waves. When you see a new wave approaching, face it, sink to the bottom, and wait for it to pass. Then push to the surface and swim forward until the next wave approaches.

Try to stay away from rocky shores. If you have to land on one, avoid the locations where waves hit high. Watch out for the white spray (an indicator of high-hitting waves). Instead, look for places where the waves roll up the rocks and approach slowly. Also, look for heavy growths of seaweed because the water will be quieter there. Instead of trying to swim through the seaweed, crawl over the top of it with overhand movements.

When you reach the shore, let a wave carry you in. Face the shore with your feet in front of you, three feet lower than your head. This way your feet will absorb the shock. Keep your hands ready to grab onto the rocks and hold on.

If you fail the first time, swim with your hands only, and as the next wave approaches assume the sitting position again with your feet facing forward.

Rescue

When you see a rescue craft approach, whether it's a ship or aircraft, immediately clear all lines and gear that could get in the way. Secure all loose items and take down canopies and sails. Fully inflate your life preserver and stay in the raft until you're instructed to leave. It's important to follow all instructions given by rescue personnel.

In the case of an unassisted helicopter rescue, first secure all loose equipment in the craft, then deploy the sea anchor. With an inflated craft, partially deflate it, then grasp the handhold and roll out. Allow the recovery device from the helicopter to reach the water's surface; while holding onto the raft handhold with one hand, grasp the recovery device with the other. Climb onto the device, secure yourself, then signal the hoist operator that you're ready to go up.

Coastal Rescue

In cases where your craft isn't sighted by rescuers, you might have to land along the coast before you're rescued. To maximize your chances of being rescued, it's better to remain close to the shore instead of going inland. But during wartime, the enemy is likely to patrol most coastlines

Coastal Health Hazards

When surviving along a coastline, you'll need to look out for coral, poisonous and aggressive fish, crocodiles, sea urchins, sea biscuits, sponges, anemones, and tides and undertow.

Coral

Coral can inflict painful cuts that have to be cleaned thoroughly to prevent infection. DO NOT use iodine to disinfect any coral cuts because some coral polyps feed on iodine, and the application of iodine can cause them to grow inside your flesh.

Poisonous Fish

Many reef fish contain poisons that are present in all parts of the fish, but especially in the liver, intestines, and eggs. Since fish toxins are water soluble, no amount of cooking will neutralize them. Also, birds are much less susceptible to the poisons. So don't think it's safe for you to eat to eat a certain type of fish just because you see a bird eat it.

Fish toxins produce numbness of the lips, tongue, toes, and tips of the fingers; severe itching, nausea, vomiting, loss of speech, dizziness, and a paralysis that can eventually result in death.

Certain species of coastal fish are dangerous to even touch. Stonefish and toadfish have venomous spines that can cause agonizing pain but are seldom fatal. Some jellyfish can inflict a very painful sting if they touch you

≫ **Life preserver**

≫ **Life vests**

≋ **Barracuda**

≋ **Jellyfish**

≋ **Crocodiles**

≋ **Sea urchin**

with their tentacles. Many stingrays have a poisonous barb in their tail. Other reef fish can deliver electric shocks.

Aggressive Fish

Avoid sharks, barracuda, sea bass, moray eels, and sea snakes. All of them will bite if disturbed. Barracuda have been known to attack swimmers and divers wearing shiny objects.

Crocodiles

Crocodiles inhabit tropical saltwater bays and mangrove-bordered estuaries and range up to sixty-five kilometers into the open sea. They're commonly found in the remote areas of the East Indies and Southeast Asia. Consider any crocodile over one meter long to be dangerous, especially females guarding their nests. Crocodile meat is an excellent source of protein, if you can trap or kill one.

Sea Urchins, Sea Biscuits, Sponges, and Anemones

These creatures—which are usually found in shallow tropical water near coral formations—can cause extreme pain but are rarely fatal. If stepped on, they slip fine needles of lime or silica into the skin, where they break off and fester. If possible, remove the spines and treat the injury for infection.

Tides and Undertow

If caught in a large wave's undertow, push off the bottom or swim to the surface and proceed shoreward in a trough between waves. Do not fight against the pull of the undertow. Swim with it or perpendicular to it until it loses strength; then swim for shore.

Coastal Food

Obtaining food along a coastline is generally not a problem. There are many types of seaweed and animal life that are easy to find and safe to eat.

Mollusks

You can eat mussels, limpets, clams, sea snails, octopuses, squids, and sea slugs. Avoid the blue-ringed octopus and cone shells. Also beware of "red tides" (algal blooms) that make mollusks poisonous.

Worms

Coastal worms are generally edible but are better used as fish bait. Avoid bristle worms that look like fuzzy caterpillars and tube worms that have sharp-edged tubes. Arrowworms, found in the sand, are edible either fresh or dried.

Crabs, Lobsters, and Barnacles

They are seldom dangerous and are excellent sources of protein. Look out for the pincers of larger crabs or lobsters, as they can crush a man's finger. Many species also have spines on their shells, so it's advisable to wear gloves when catching them. Barnacles can cause scrapes or cuts and are difficult to detach from their anchor, but the larger species especially are an excellent food source.

Sea Urchins

Although they're a good source of food, they can cause painful injuries when stepped on or touched. Handle them with gloves, and remove all spines.

Sea Cucumbers

These are an important food source in the Indo-Pacific regions. Remove the five muscular strips that run the length of the body, and eat them smoked, pickled, or cooked.

Sea Survival Stories

San Blas Fishermen

Four Mexican fishermen—Salvador Ordonez, Jesus Vindaña, Lucio Rendón, and a man called El Farsero—and the owner of a small fishing boat, Juan David, left the fishing village of San Blas on October 28, 2005, on a shark fishing expedition. Heavy currents caught the boat and carried it nearly 5,000 miles out to sea. Three of the five men survived and claimed that they stayed alive by catching and eating fish and seabirds. They weathered storms, and staved off thirst by collecting rainwater. A Taiwanese fishing boat rescued the men on August 9, 2006, near the Marshall Islands, nine months and nine days after they left shore, making their feat of survival one of the longest on record.

Mr. Ordóñez, a thirty-seven-year-old native of Oaxaca who was known in San Blas for carrying a Bible everywhere and getting into the

⌃ **Mussels**

⌃ **Cuttlefish**

⌄ **Barnacles**

occasional bar fight, said he worried at the start of the trip because the boat's owner had not packed enough provisions for the usual shark-fishing trip, which lasts from three days to a week.

Once disaster struck and the men found themselves being sucked out into the Pacific Ocean, they turned to what they knew best: fishing. Having some knives and other equipment aboard, they fashioned hooks from engine parts and lines from cables. Mr. Ordóñez, who had taken a course on surviving at sea a year before, earned the nickname "the cat" for his ability to sneak up on seabirds.

The men survived on fish, birds, and fish blood. But the boat's owner, Juan David, and El Farscero had trouble digesting the raw food, and became sick and started vomiting blood. They both died after three months and, according to the three survivors, had to be thrown overboard. As they drifted for months Ordonez, Vindaña, and Rendón passed the time singing ballads, dancing, pretending to play guitar, and reading aloud to each other from the Bible. Little things took on tremendous importance. They kept track of time with Mr. Rendón's wristwatch.

The worst part of the voyage came in December and January, when several large storms hit and the men could catch few fish. "We were afraid we would sink," Mr. Vidaña said. "The longest we went without food was about thirteen days, when we had only one seabird to eat."

Poon Lim

Born in China in 1917, Poon Lim was working as a second steward on the British merchant ship *SS Ben Lomond* when it was intercepted and torpedoed by a German U-boat on November 23, 1942. As the ship was sinking in the Atlantic Ocean off the southern coast of Africa, Poon Lim grabbed a life jacket and jumped overboard before the ship's boilers exploded. After approximately two hours in the water, he noticed an empty life raft and climbed into it. The raft contained a couple of tins of biscuits, a ten-gallon jug of water, some flares, and an electric torch.

Poon Lim kept himself alive by drinking only a few swallows of water and taking two biscuits a day. He kept himself in shape by swimming around the raft twice a day. He took a wire from the electric torch and made it into a fish hook and used a hemp rope as a fishing line. When he captured a fish, he cut it open with the edge of a biscuit tin and used half of it for bait. Later he captured seagulls for food. He used the canvas of the life jacket to collect rainwater.

Twice other vessels passed nearby—a freighter and U.S. Navy patrol planes—but they didn't notice him. Poon Lim counted the days with notches on the side of the raft. On April 5, 1943, he saw a sail on the horizon and managed to attract attention. Three Brazilian fishermen rescued him and took him to Belem three days later.

During his four-month ordeal Poon Lim had lost twenty pounds, but he was able to walk unaided. He later found out that only eleven of the ship's 55-man crew had been rescued. King George VI honored him with a British Empire Medal and the British Navy used his tale in its manuals of survival techniques. His employers gave him a gold watch.

⌃ *Titanic* **lifeboats on way to Carpathia**

The *Titanic*

The "unsinkable" *RMS Titanic* was the biggest and most luxurious passenger liner of her time. On April 10, 1912, the *Titanic* left England on its maiden voyage with about 2,200 people onboard, including many of the world's rich and famous. Four days later, a lookout spotted an iceberg directly ahead. Titanic reversed her engines and tried to turn away, but it was too late. The *Titanic* hit the iceberg, which gouged out 250 feet (eighty-three yards) of hull and popped out at least six rivets below the waterline. Water poured through the holes from the rivets and flooded the first five watertight compartments at the front of the ship. The weight of the flooded compartments pulled *Titanic* headfirst into the ocean. Over the next two hours, the *Titanic* broke into at least two pieces and sunk.

Even though she carried more lifeboats than were required by law, there were only enough to accommodate about half of the passengers onboard. And in the panic, most of the available lifeboats were launched only half or a quarter full. Only two out of the eighteen launched lifeboats returned to rescue passengers from the water after the wreck.

More than 1,500 people died. Many went down with the ship or died of hypothermia after jumping overboard into the frigid water. There was another ship in the vicinity, but her wireless operator had already left his post for the night, so he didn't pick up *Titanic's* calls for help. Fewer than 700 people survived.

The Whale ship *Essex*

In 1820, when the *Essex* set sail from Nantucket Island, off Rhode Island, on a routine whale hunt, the crew of twenty men had no idea that fifteen months later they would find themselves adrift in the vast Pacific at the mercy of the elements and their own human failings. But like the classic novel it inspired—Herman Melville's *Moby Dick*—the *Essex* was hit by an enraged sperm whale that struck just beneath the anchor.

⌃ **Vessels prepared for water rescue**

According to the account of crew member Owen Chase: "I turned around and saw him about one hundred rods (550 yards) directly ahead of us, coming down with twice his ordinary speed [around twenty-four knots or forty-four kph], and it appeared with tenfold fury and vengence in his aspect. The surf flew in all directions about him with the continual violent thrashing of his tail. His head about half out of the water, and in that way he came upon us, and again struck the ship."

The whale crushed the bow and drove the 283-ton vessel backwards into the ocean. Then it disengaged its head from the shattered timbers and swam off, never to be seen again.

Able to salvage only three small harpoon boats and a few meager supplies, the crew made the fateful choice to sail back east to South America rather than west to any of several Pacific islands. With an innate fear of cannibalism rumored to exist among native tribes, they preferred to brave the familiarity of the ocean. For the next ninety-three days they would come to question the wisdom of that decision as the trade winds and storms blew them farther and farther from their destination. The sad irony is that many times during their 3,000-mile ordeal of hunger, thirst, and death they would have been within reach of a lifesaving island if they had only turned west.

As members of the crew began to die of thirst, the three harpoon boats landed on uninhabited Henderson Island—roughly 1,350 miles south-southeast of Tahiti and 4,100 miles west of Panama—within the British territory of the Pitcairn Islands. There the men found fresh water and gorged themselves on birds, fish, and vegetation. Within a week they had exhausted the island's resources.

Three men—William Wright, Seth Weeks, and Thomas Chapple—opted to stay behind on Henderson. The remaining *Essex* crewmen resumed their journey, but within three days had exhausted the fish and birds they had collected for the voyage, leaving only a small reserve of bread salvaged from the *Essex*. One by one, the men began to die.

By the time the last of the eight survivors were rescued on April 5, 1821, the corpses of seven fellow sailors had been consumed. All eight returned to the sea within months of their return to Nantucket.

7

BASIC SURVIVAL TIPS

"If hope is out there, hope can get you through."

—*NASA Astronaut Jerry Linenger, who spent months stranded on the Russian Space Station*

Main Tips to Remember

The only person you can ultimately count on is yourself.

No one else can give you the mental will, physical stamina, and common sense that you're going to need to survive. So don't depend on others—you may be alone! Make your plans, pack your own survival kit, and if something unplanned happens when you are on your own in the wild, be prepared to take care of your own needs as well as the needs of your teammates.

This self-sufficient attitude is empowering in itself. Remember that your life depends on what you do, not on the chance that a teammate will be there to do for you what you can't do for yourself.

Always leave behind detailed plans and timetables with a trusted person.

That way, if you're missing, a search party is likely to be sent out sooner than later if you fail to arrive back when expected.

Prepare for the six contingencies.

1. Becoming lost. It's not enough to rely on your good sense of direction. Always carry at least one compass, a map, and GPS and a SPOT or locating device.

2. Darkness. With darkness we shift from relying primarily on seeing to relying primarily on hearing. This is an uncomfortable change for some people. Remember that darkness can be your friend. Treat it with respect, and don't move into areas where you could get hurt by your inability to see.

3. Being stranded. There are countless contingencies under which you could be stuck in the wilderness for an extended period of time. Anticipate that this could happen and plan for ways to alert others and make your way to safety.

4. Illness or injury. Any time you go into the wilderness, there's always the possibility that you can become injured or ill. Practice and develop your own wilderness first-aid skills.

5. Extreme weather. There is no such thing as bad weather, just different types of weather. Always be prepared. Snow, rain, or extreme heat or cold can impact your ability to survive. Before heading out, make sure you have the proper clothing, water, and the ability to shelter yourself for extended periods of time.

If lost, stop and do a map study.

Don't move unless you know where you are and where you are going. Many very experienced point men and navigators have become lost or disoriented in the wild. Remember that the consequences of panic can be fatal. Take a break, do a good map study, reevaluate your situation, and allow the adrenaline that has flooded your system and put you in fight or flight mode to subside.

≈ **Maps**

Assess your situation as objectively as you can.

1. Treat any injuries—yours or your teammates. Self aid and buddy aid. Your health is most important for survival.
2. What needs to be done to assure your safety? Do you need to move to a safer area?
3. Observe the area of your location. What are the hazards? Are there enemy or friendly forces in the immediate area? What are the advantages? Is there water nearby? What can you take advantage of to help you survive?
4. Plan your next move carefully. Work out a plan in your head first. If you're satisfied with it, proceed. If not, give yourself time to come up with a better alternative.

Take stock of your supplies and immediate needs.

A healthy man can survive for several weeks without food and several days without water. So water is your most important requirement. Under normal circumstances, the human body requires two quarts of water daily to maintain adequate hydration.

Don't ration the water you have to last for many days. Drink what you need. It's better to have water in your body than in a bottle or canteen. Conserve water lost through sweating by wearing a hat, sitting in the shade, moving only at night, and so on.

In most terrain, you will eventually find water when moving downhill. Watch animals, or follow their tracks. They'll usually lead to water. Birds tend to congregate near water, too. Remember that water from streams and ponds should be boiled before drinking or purified by other means—tablets, straw, filter, boiling etc.

Signal.

Always carry a whistle, mirror, and matches to start a fire. Smoke is visible from far away in the day.

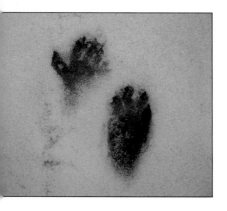

⋩ Raccoon footprints

Find food.

Food isn't an immediate concern unless you're reasonably sure that rescue is many days or weeks off. As a general rule, avoid plant life unless you know for a fact that something is edible. If it walks, swims, flies, slithers, or crawls, it's probably safe to eat. All fur-bearing animals are edible. All birds are edible. Grubs found in rotten logs are edible, as are almost all insects.

Fire requires three elements:

Oxygen, fuel, and a source of heat. Is your fuel thin and dry enough? Is your heat source hot enough to light the tinder? Is there enough oxygen reaching the point where the heat meets the fuel? Identify the problem and proceed.

Survival is the ability and the desire to stay alive, sometimes alone and under adverse circumstances.

Understand and master each part of this definition.

1. Ability. Be proficient at building shelter, starting a fire, signaling for help, and staying hydrated.
2. Desire. Regardless of how bad the situation might be, never lose the will to survive and always maintain a positive attitude.
3. Stay alive. Your ability to effectively deal with life-threatening medical situations is of the highest priority. Stay current with your emergency medical skills.
4. Under adverse conditions. The more you know about your environment ahead of time, the greater your advantage.
5. Alone. Never count on the help of others. Be self-proficient since you may end up alone.
6. Until rescued, be patient. It's your job to keep yourself and your teammates alive.

Learn to deal with the enemies of survival.

1. Pain
2. Cold and/or Heat
3. Thirst
4. Hunger
5. Fatigue

The Rule of Threes

A human being can survive:

- three minutes without air
- three hours without a regulated body temperature
- three days without water
- three weeks without food

In summary, here is a list of common-sense survival tips:

1. Always carry a map, at least one compass, a GPS and a SPOT or something similar.
2. Dress using layers to avoid overheating.
3. Carry water and water purification tablets or filter when possible. During the map studies, always identify alternate water sources.
4. Carry high-calorie energy food such as protein bars in your second and third line gear.

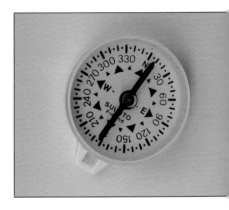

⌃ **Compass**

The Case of Aron Ralston

Can you imagine amputating your arm with a blunt knife? As excruciatingly painful and inconceivable as it sounds, that turned out to be the only option left to 28-year-old Aron Ralston after an 800-pound boulder fell on his arm, pinning it to a canyon wall.

From midday Saturday, April 26, 2003, until midday Thursday, May 1, Ralston was stuck in a remote area of Canyonlands National Park in Utah alone and unable to free himself. He had little food and water. No one would even wonder where he was until he didn't show up for work on Tuesday. Unable to sit, lie down, use his right arm, or sleep, he knew that he was in for an excruciatingly difficult time. Those 120 hours of what he calls "uninterrupted experience" tested to the fullest his physical, mental, emotional, and spiritual being.

Finally, on May 1, 2003, he did the unthinkable, first using the boulder to leverage his arm until the bones snapped and then sawing away at muscle and tendon with his pocketknife. He then rappelled down a sixty-five-foot wall and was later found by hikers as he walked back to his car.

He survived, wrote a best-selling book about his experience (*Between a Rock and a Hard Place*), and continues to climb. He later admitted that his big mistake was not telling anyone where he was going.

⌄ **Canyonlands National Park, Utah**

8

WATER

"Water is the driving force of all nature."

—*Leonardo Da Vinci*

Getting lost or stranded in the wild is something that can happen to anyone, whether you're a Navy SEAL, experienced outdoorsmen, hiker, tourist, or just someone out for a weekend drive. Anybody can be forced to deal with circumstances beyond their control, alone and lost, with only their wits to rely on for survival.

The human body is composed of up to seventy-eight percent water. So it's no surprise that the single-most important thing you need to live is not food; it's water. The good news is that if you're resourceful and know where to look, you can find or collect good drinking water in just about any environment on earth.

The Institute of Medicine currently recommends a daily intake of approximately 2 to 2 ½ quarts of water to replace the water lost through normal body functions—urination, defecation, breathing, and sweating. All of the chemical and electrical activities that take place in the human body take place in a water environment; when water is in short supply, these activities begin to malfunction.

It's important to understand that many people begin their survival already dehydrated due to stress and other factors. They often continue to dehydrate further when water supplies are limited and the quality of any available water is suspect. People needing water, but fearful that it is contaminated with *Giardia, Cryptosporidium,* or other harmful pathogens, often put off drinking or choose not to use the water at all.

In North America, as a general rule, it is better to drink available fresh water.

⩔ **Pond**

If the water contains harmful pathogens, the onset of symptoms will usually be days, if not weeks away. By then the individual will hopefully have access to medical care.

The one exception to this rule is that certain lakes mainly found in the western United States contain high concentrations of calcium carbonate and calcium bicarbonate. This water is not potable. Lakes containing these substances are usually easy to identify because the calcium salts leached from the soil are deposited in the form of white powder around the shorelines. This water tastes terrible and should not be consumed unless there is absolutely no other water source available.

In other parts of the world, especially developing countries, drinking water that has not been disinfected is NOT recommended.

Viruses such as hepatitis, not commonly found in North American waters, are prevalent here and can quickly cause incapacitating illness.

Finding Water

Throughout much of North America, fresh water can usually be found in open sources such as lakes, ponds, rivers, and streams. In most cases, it can be obtained fairly easily. Remember that water always seeks the lowest level possible and that, if present, some form of vegetation will most likely grow nearby.

The best way to locate water is from a vantage point that allows you to scan the surrounding countryside. Slowly and methodically look for indicators such as green vegetation, flocks of birds, trails left by domestic and wild animals, and even large formations of rock that can contain natural springs. Check for low-lying areas—such as depressions or sinks—where rainfall or melting snow is likely to collect. Water can often be found in these areas long after the last precipitation, especially if they are shaded.

Water sources like these should be checked carefully since they're often contaminated with debris that has been washed into the drainage. Finding the remains of animals that have died nearby or in the water and other similar contaminants will necessitate boiling the water, the use of halogens (iodine or chlorine), or the use of a mechanical purification pump.

The quantity of water produced by seeps and springs tends to vary greatly. Some of them produce no more than a few teaspoons of water per hour. In other cases, gallons of water can flow from the ground in minutes. Where the output is slow and small, use the flat edge of the mouth on a plastic bag to scoop up the water from a shallow source; if it is flowing, use it to collect the water as it runs into the bag. A short piece of vinyl aquarium hose also works well for sucking up water from shallow collections or to recover water from narrow cracks in the rocks.

Also, keep an eye out for man-made sources of water such as windmills, wells, tanks, dams, and irrigation canals. Windmills are common in parts of North America, especially in areas where little surface water exists. In most cases, the water pumped to the surface

⩣ **Mountain lake**

is collected in a nearby tank or pumped directly into a trough from which livestock can drink. Where an open source is not available, it may be necessary to dismantle the piping associated with the windmill to gain access to the water.

If you find an abandoned well where the rope and bucket typically used to lift water from these wells is missing, improvise a means to lower a container down into the well to retrieve the water. If you don't have a container, an item of clothing can be lowered into the water to serve as a sponge.

In arid areas, particularly in the western and southwestern United States, many state wildlife agencies and conservation organizations have installed rainwater collectors called "guzzlers." These are designed to gather precipitation and feed it into a holding tank, where it remains until it is either consumed by animals or evaporates.

Just because there's no water visible on the surface of the ground, that doesn't mean that it's not present in the soil in sufficient quantity to be collected. Locate low-lying areas where water is most likely to have accumulated and dig down until damp layers of soil are found. The hole should be about a foot in diameter. Over time, water may seep into the hole where it can be collected. If no indicators of subsurface water are present, dig a hole in the outside bend of a dry riverbed. Look for a location where the centrifugal force of flowing water has eroded the outer bend, creating a depression where the last remnants of water flowing downriver will have accumulated.

Groundwater collected this way is likely to be muddy, but straining it through cloth will clean it and will get you by in the short term. It's important to remember that you're taking a risk anytime you drink ground water without purifying it.

Rain is a great source of drinking water and in most rural areas can be consumed without risk of disease or illness. If you have a poncho or some plastic sheeting, spread it out and tie the corners to trees a few feet off the ground. Find a container and tie the plastic on a slant so that the rainwater can drain into it. If you can't find a container, devise a makeshift water bag by tying the plastic level on all four corners but letting it sag in the middle so that the rainwater can collect there. If the rainwater tastes different than what you're used to, it's because it lacks the minerals that are found in groundwater and in streams. If you don't have a poncho, rain gear, or piece of plastic, remember that water will collect on the upper surfaces of any material (it doesn't have to be waterproof) and drain to the lowest point, where it can be collected in a bucket or other container.

Melt snow before you consume it because if you eat it frozen, you'll reduce your body temperature, which can lead to dehydration. The best technique to convert snow into water is by using what military survival schools call a water machine. Make a bag out of any available porous fabric (you can use a T-shirt), fill it with snow, and hang it near (but not directly over) a fire. Place a container under the bag to collect the water. By continually filling the bag with snow you'll keep it from burning.

If your circumstances don't allow you to make a fire, you can melt snow with the heat of your body. But the process is slow. Put several cups of snow in any available waterproof container (preferably a soft plastic water bag, locking sandwich bag, or something similar) and place it between layers of your clothing or in your sleeping bag. Since the amount of heat needed to convert snow to water is large and the amount of body heat available is finite, only small quantities can be melted at a time.

⌃ **Water droplets collecting on leaf**

Collecting Water

Heavy dew can be a good source of potable water. Before the sun rises, tie absorbent cloth around your shins and walk through high grass. This way you might be able to collect enough water for an early morning drink.

Bamboo

Fruits, coconuts, cacti, vines, palm trees, and bamboo can also be good sources of liquid sustenance. Bend the top of a green bamboo stalk down about a foot off the ground and tie it off. Cut a few inches off the tip, put a container underneath, and leave it overnight. The next day, you're likely to find a nice amount of clear, drinkable water.

Vines

Water-producing vines varying in size from the diameter of a pencil up to the thickness of a man's forearm can be found throughout much of the southeastern United States. The thicker the vine, the more water it is capable of producing. Select the thickest one first.

Use a sharp knife or a machete to sever the tough, woody vine. Vines that exude a white latex sap or those that produce a colored or foul-smelling sap should be avoided. If no sap is observed, or if the sap that is observed is clear and without aroma, remove a twenty-four-inch section, severing the higher end first and then the lower end. If the lower end is cut first, the water contained within the vine is drawn up by capillary action and far less water will drain out by the time the upper end is severed.

⌃ **Drinking water out of bamboo**

⌄ **Bamboo**

⚑ **Drinking water out of a vine**

⚑ **Tropical vines**

⚑ **Cactus**

Once removed, hold the section of vine vertically and the water in it will drain into a container (or a cupped hand), where it should be evaluated. Any liquid that is colored should not be consumed. Liquid that has an unpleasant aroma other than a faint "woody" smell should not be consumed but can be used to satisfy any hygiene needs. Taste a small amount of the water. Water that has a disagreeable flavor other than a slightly "earthy" or "woody" taste should not be utilized for drinking. Hold a small amount of water in your mouth for a few moments to determine if there is any burning or other disagreeable sensation. If any irritation occurs, the water should be discarded. Liquid that looks like water, smells like water, and tastes like water is water and can be safely consumed in large quantities without further purification.

Cactus

Cactus as a source of water is often overrated. But if you decide to approach one, use caution, as the thorns usually cause infections. Use sharp sticks or knifes to handle cactus safely. Any injury from a cactus plant should be treated immediately to reduce the risk of infection.

Although all cacti can be used for gaining additional moisture, it can take a great deal of work to open a full-sized barrow cactus and fight with the spiny thorns that protect it. If you decide to take on a cactus, do it in the cool of the evening. Using caution, remove the top of the barrow cactus. Once the top is off, you will find a white substance that resembles watermelon meat inside (this is a liquid-filled inner tissue). Using your knife, cut out hand-size chunks and squeeze the moisture from them.

Prickly pears are easier to collect and prepare. Use a large sharp stick and a good knife. Stab the round prickly pear with the stick, and then cut it off with the knife. Next, use a fire to burn the thorns off of the cactus. Make sure you sear the cactus well to remove even the smallest thorns.

Once the thorns are removed, peel the green- or purple-colored outer substance off, and eat the inside. Prickly pear meat tastes so good that in Arizona and New Mexico people make jellies and candies from it. Chew the moisture-filled inner tissue, not the rough outer "bark."

Getting Water from Plants

The use of clear plastic bags to enclose living vegetation and capture the moisture transpired by the leaves can be an effective method of collecting water. A plant's survival is dependent on its ability to gather water from the soil. This water is passed up through the plant's roots, stems, and branches, and is finally released back to the atmosphere through pores in the leaves as water vapor—a process called evapotranspiration.

This water vapor can be collected with a clear plastic bag. It works best when the vegetation is high enough to be off the ground. Shake the vegetation to remove any insects, bird droppings, or other materials that might contaminate the water. Insert the limb or bush just like you would a hand into a mitten. Then, tie the open end of the bag around the tree or bush and seal the opening shut with a cord or duct tape. At the closed end of the bag, tie a rock so the bag is weighted and forms a collection point for the water.

Within a short period of time, water will begin to condense on the inner surface of the bag, collect into water droplets, and drain to the lowest point of the bag. The quantity of water obtained in this manner is dependent on the amount of water in the ground and the type of vegetation used. Other factors that will determine water production include the amount of sunlight available (it doesn't work at night), the clarity of the plastic bag, and the length of time the process is allowed to work. It is not uncommon to find that two or three cups of water, and sometimes much more, have accumulated over a six- to eight-hour daylight period.

The best way to remove the water without disturbing the bag is to insert a length of vinyl aquarium hose through the neck of the bag down to the lowest point where water will collect. The water can then be sucked out or siphoned into a container. When enclosing vegetation in the plastic bag, it is advisable to place a small stone in the lower corner where the water will collect. The weight of the stone creates a separation between the enclosed plant life and the water and will keep plant saps from contaminating the water.

Similarly, leaves and small branches can be cut and placed in a clear plastic bag. In this method, heat from the sun causes the liquids in the foliage to be extracted and collect in the bag. However, this method may produce water containing unsafe toxins. Taste it first. If the water is bitter, do not drink it.

Solar Stills

The quantity of water produced by a solar still depends on the amount of water contained in the ground. Because of this, solar stills are not reliable for obtaining water in arid areas since desert soils tend to hold little or no water. The amount that a survivor is likely to obtain via this method must be balanced against the amount of sweat lost while constructing the device.

⌃ **Solar still**

However, in other types of climates, a solar still can be very effective way of capturing water.

To build a solar still, dig a hole approximately one meter across and two feet deep. Dig a smaller hole, or slump, in the middle of the hole. Place a container in the slump to collect the water. Then, cover the hole with a plastic sheet and secure the edges of the sheet with sand and rocks. Finally, place a rock in the center of the sheet, so it sags.

During daylight hours the temperature in the hole will rise due to the heat of the sun, thereby creating heat vapors which will condensate on the inside of the plastic sheet and run down. It then drops into the container in the sump hole.

You should never drink the following:

- Blood
- Urine
- Saltwater
- Alcohol
- Fresh sea ice

Fresh sea ice is milky or grey, has sharp edges, does not break easily, and is extremely salty. Older sea ice is usually salt-free, has a blue or black tint and rounded edges, and breaks easily. Melted old sea ice is usually safe to drink, but should be purified first, if possible.

Waterborne Contaminants

In most parts of the world, surface water is seldom pure. There are five basic waterborne contaminants that you should be particularly aware of: turbidity, toxic chemicals, bacteria, viruses, and parasitic worms.

Turbidity

A measure of the cloudiness of water, or more specifically a measure of the extent to which the intensity of light passing through water is reduced by suspended matter in the water. The sources of turbidity can be attributable to suspended and colloidal material, and may be caused by several factors such as: microorganisms and organic detritus, silica and other sands and substances including zinc, iron and manganese compounds, clay or silt, the result of natural processes of erosion and/or as waste from various industries.

Toxic Chemicals

Dangerous and toxic chemicals include, among others, pesticides, herbicides, fertilizers from agricultural land and runoff from household and industrial chemicals.

Bacteria, Viruses, Parasitic Worms

Giardia lamblia is a parasite that lives in the intestines of humans and animals. It's expelled from the body in feces, and is found worldwide and in every region of the United States. It causes giardiasis, which produces cramping, nausea, and diarrhea. Symptoms may not show up for two weeks, and once present can last as long as six weeks. If infected, get medical attention as soon as possible.

Cryptosporidiosis is another waterborne illness caused by parasites found in feces. The same symptoms as giardiasis can be expected, but more severe. Both of these parasites can be found in soil and vegetation as well, so wash anything you plan on eating in purified water and remember: To give yourself the best chance at survival, always boil your water, even if it looks clean.

⌃ **Giardia**

Water Purification and Disinfection

To be safe to drink, water must be disinfected so that all harmful microorganisms are removed. To do this water must be boiled, treated with chemicals, or filtered. "Disinfection" of water should not be confused with "purification" of water. Some of the methods used to purify water may not remove or kill enough of the pathogens to ensure your safety. Make sure the water you drink is disinfected.

The first step to disinfecting water is to select the cleanest, clearest source of water available. Inorganic and organic materials such as clay, silt, plankton, plant debris, and other microscopic organisms will reduce the effectiveness of either chemical or filtration disinfection. Chemicals used to disinfect water will clump to any particulate in the water, thus reducing its ability to disinfect the water. And water containing a lot of material will quickly clog a filtration system. For the best results, collect water from below the surface but not off the bottom. When collecting murky water, allow it to settle and then filter it through your shirttail, bandanna, or other piece of cloth.

Remember:

Filtering water doesn't always purify it, but it does reduce particles and sediment and make the water taste better. However, there are microbial

⌃ Boiling water

purification filters on the market that not only remove parasites such as Giardia, but also kill waterborne bacteria and viruses. These types of filters are optimal.

- **Boiling** is the best way of killing all microorganisms. Boiling will not neutralize chemical pollutants.
- To purify water with **chemicals**, use water purification tablets.

Boiling

Bringing water to a boil kills any organisms in it. In most cases, water does not have to be boiled for a specific length of time. The time it takes to bring water to a boil and the temperature of the water when it boils is sufficient to kill *Giardia*, *Cryptosporidium*, and any other waterborne pathogens. While the boiling point of water decreases as you climb higher, the temperature at which the water boils is still hot enough to kill those organisms that might make you sick. Continuing to boil the water wastes fuel, evaporates the water, and delays consumption.

Overseas, especially in developing countries where river systems are still a frequent method of sewage disposal, boiling for a longer period of time (one or two minutes) is advisable.

Chemical Purification

Chemicals that have the ability to disinfect water are known as halogens, and include iodine and chlorine. The effectiveness of halogens is directly related to their concentration, the amount of time they are left in contact with the water, and the temperature of the water—the colder the water the longer the contact time.

Iodine

Comes in tablet and liquid forms. I recommend the tablets because liquid iodine is messy and the containers are prone to leaking. Potable Aqua tablets (which contain iodine) are used by the U.S. military and many disaster relief agencies.

Iodine kills harmful bacteria, viruses, and most protozoan cysts found in untreated water. (It is NOT effective on *Cryptosporidium*.) The recommend dosage of two tablets per quart or liter of water is sufficient to kill organisms such as *Giardia*. Once the tablets are placed in the water, they should be allowed to sit for at least thirty minutes (even longer if the water is very cold), and then shaken so that the iodine and water mix thoroughly. The dissolved tablets will leave a slight iodine taste in the water, which some find disagreeable. Lemon juice, lemonade, Kool-Aid, or Gatorade powder can be added to neutralize the iodine flavor.

Iodine tablets are commonly packaged with a second bottle of ascorbic acid (PA Plus) tablets that deactivate the iodine, making the water pleasant to drink. One tablet is usually enough to reduce the iodine taste.

Iodine tablets deteriorate on exposure to heat, humidity, or moisture. Over time, opening and closing the cap to remove tablets results in the normally gray-colored tablets changing to green or yellow. Once they have changed color, they have lost their effectiveness and shouldn't be used. Avoid using the military iodine tablets that are sometimes found in military surplus stores. The military got rid of them because their shelf life has expired.

Advantage of iodine tablets:

- Easy to use
- Lightweight
- Inexpensive

Disadvantages:

- Not effective against *Cryptosporidium* cysts
- Some people are allergic to iodine
- People with known thyroid problems should not use iodine
- Iodine should not be used as a long-term (more than six weeks) method of purifying water due to its potential harmful effects on the thyroid.

Chlorine

An effective agent against bacteria, viruses, and, unlike iodine, cysts such as *Cryptosporidium*. Another advantage of using chlorine is that it leaves no aftertaste. On the downside, a significant disadvantage of using chlorine tablets is that you have to wait for four hours after adding a tablet before you can drink the water.

Advantages of Chlorine tablets:

- No aftertaste
- Chlorine kills *Cryptosporidium*

Disadvantages:

- Four-hour contact time

Almost all laundry bleaches, including Clorox, contain five and one-half percent sodium hypoclorite, which is a suitable purification chemical

⌃ LifeStraw

for water. Put a small amount in a bottle with an eyedropper dispenser and add it to your E&E kit. Make sure you do not use powdered, scented, or other non-pure bleaches.

Before adding bleach to the water you want to purify, remove all suspended material by filtration (through a cotton cloth, improvised sand filter, or other means) or by simply allowing sediment to settle to the bottom.

Add eight drops of bleach per gallon of water (or two drops per quart). If the water was filtered, then shake it up to evenly dispense the bleach, and wait fifteen minutes. If the water has sediment on the bottom, don't shake it up. Instead, allow the treated water to stand for thirty minutes.

Because killing microorganisms also consumes the bleach, you can tell by smelling whether or not there's anything left to kill. If there's no chlorine odor then all of the bleach was used up, meaning there could still be living organisms. Repeat the dosage and allow the water to stand for another fifteen minutes. If there is any chlorine odor, however faint, after thirty minutes, all of the bacteria, viruses, and other microorganisms are dead, and the bleach has done its job with some to spare.

When treating cloudy, green, or really nasty water (swamp water, for example), start with sixteen drops of bleach per gallon of water (or four drops per quart). Smell the water. If there's a faint odor of chlorine, the water is drinkable. If not, then repeat the treatment.

Treating Larger Quantities of Water

A teaspoon of bleach treats about 7 ½ gallons of clear water or about four gallons of dirty water. Therefore, a tablespoon of bleach treats about twenty gallons of clear water or about ten gallons of dirty water. A quarter cup of bleach will purify about ninety gallons of clear water or forty-five gallons of dirty water.

LifeStraw

The LifeStraw is a portable filtration device that enables you to safely drink directly from any fresh water source. The straw itself is about eleven inches long, less than one inch around, and looks like a jumbo drinking straw. One end has the narrow mouthpiece; the other goes directly into the water source. Each LifeStraw lasts for 185 gallons, roughly the amount of water needed for one person per year.

The filter is designed to eliminate 100 percent of waterborne bacteria, almost ninety-nine percent of viruses, and particles as small as fifteen microns.

9

SHELTER AND FIRE

"On the occasion of every accident that befalls you, remember to turn to yourself and inquire what power you have for turning it to use."

—*Epictetus*

Shelters

A healthy human can survive for several weeks without food and several days without water, but in many cases only several hours without proper shelter from the elements. After water, shelter is a critical need in terms of your survival.

From keeping you protected from the elements to providing a place to rest, wilderness shelters serve a key role in survival situations. Not only do they provide for physical needs, but they also help create a sense of home in the wilderness.

Your environment and the equipment you carry with you will determine the type of shelter you're able to build. Before you head out, evaluate the weather in the area you're going to be in and to what extremes it is likely to reach. Shelters can be built in wooded areas, barren plains, jungles, deserts, and snow-covered mountains. Several items found in a decent survival kit—such as a Tyvex and a survival blanket—will help immensely. Wooded areas provide timber for shelter construction, fuel for fire, concealment from observation, and protection from the wind.

Be careful not to damage your waterproof gear. Instead of poking holes in a tarp to tie it off, push a small pebble up from under the tarp, and tie off around it. Try using rocks instead of stakes to hold down corners.

Size and Location

The key to making a successful wilderness shelter is choosing a good location. In most cases, higher elevations are exposed to much more wind than valleys and lower areas. And don't be fooled by the air temperature. While a thermometer might indicate that the air temperature has increased several degrees by moving to higher ground, the temperature as far as your body is concerned is likely to have dropped twenty to thirty degrees because of the windchill factor.

A good location for building a wilderness shelter is one that 1) provides easy access to ample building materials, including dead sticks, leaves, and grasses; and is 2) away from major hazards such falling branches, avalanche zones, pooling water, and insect nests. You also want a location that has a large enough flat area to allow you to lie down and sleep comfortably.

A common mistake people make when building wilderness survival shelters is that they make them too large. Not only does this require more materials, effort, and time to construct, but it often ends up being cold due to the amount of space on the inside. As a general rule, build your shelter so that it's not much bigger than you are and gives you just enough room to lie down. This will allow your body heat to keep it warmer than the ambient temperature.

All shelters should be constructed with safety in mind. Large, strong branches can provide the initial framework for many types of survival shelters. Typically, they should be strong enough to easily support the weight of an adult. This is especially important when constructing lean-to and debris hut-style shelters.

Whether you're in a hot and sunny environment, a cold and wet forest, or on a snow-covered mountain, insulation and cover can keep you protected from the elements. Use leaves, grasses, small sticks, ferns, and pine needles to provide insulation. Pile on as much as you can. Add bark or soil to the top and sides of your shelter to create a barrier from cold wind and rain.

In cool and cold environments, the primary purpose of your shelter should be to help you stay warm and avoid hypothermia. With wilderness survival shelters, there are typically two heat sources: your body heat (and that from others with you), or fire. Wilderness shelters that rely on body heat as the primary heat source (such as a debris hut), need to be small on the inside and have lots of extra insulating debris. If you plan on using a fire inside your shelter, carefully plan how to tend it throughout night, and be sure to collect a full night's worth of firewood before last light.

Never sleep directly on the ground. Make a mattress from pine boughs, grass, leaves, or other insulating material to keep the ground from absorbing your body heat. And never fall asleep before turning off your lamp or stove.

Carbon Monoxide Poisoning

Carbon monoxide poisoning is colorful, odorless, can be deadly, and can result from a fire burning in an unventilated shelter. Remember that any open flame will generate carbon monoxide. Always double-check your ventilation. Even in a ventilated shelter, smoldering embers can cause carbon monoxide poisoning. In most cases, there are no symptoms. Unconsciousness and death can occur without warning. Sometimes, however, pressure at the temples, burning of the eyes, headache, pounding pulse, drowsiness, or nausea may occur. The one characteristic, visible sign of carbon monoxide poisoning is a cherry-red coloring in the tissues of the lips, mouth, and inside of the eyelids. Get into fresh air at once if you experience any of these symptoms.

≫ **Tarp shelter**

≫ **Bamboo hut**

Shelter Types

There are many types of field-expedient shelters to consider, including natural ones such as caves, hollow stumps and logs, and those that are relatively easy to build, including debris huts, lean-tos, scout pits, and snow shelters.

Debris Huts

These are often the easiest and most practical types of shelter to build in any type of environment.

≫ **Grass and log hut**

1. Find a long, sturdy pole to serve as the main beam of your hut. It should be 1 ½ to 2 times your height. You can also use a fallen tree. Look for something—a rock, tree stump, or tree with a forked branch—that's strong enough to hold the main beam off the ground. It should be a little taller than you are when sitting.

2. Lean smaller poles or branches against both sides of your main beam at a forty-five-degree angle. Place them close together and fill in the space between them with smaller twigs and branches.

3. Cover this framework with dead leaves, dry ferns, evergreen branches, grass, or whatever other materials you can find. Keep piling on insulation materials until the sides of your shelter are at least three feet thick.

4. Place another layer of small, light branches over the outside of the hut to prevent your insulation from blowing away.

5. Put a one-foot layer of debris inside your shelter, and choose materials that will be soft enough to sleep on.

6. At the entrance to your shelter, which should face away from the prevailing wind, build a pile of insulating material that you can drag in

place once you are inside and can serve as a door. You can also build a door out of other types of material in the vicinity. You can, for example, make a door by gathering finger-sized pieces of wood and lashing them into a grid pattern. Make two grids, then place debris between them. Lash the grids together and you have an insulated door.

Lean-Tos

These are easy to build, suitable to most types of terrain, and should always be built with their back toward the wind.

1. Insert two Y-shaped sticks into the ground, about one foot deep. Each should stand about three feet high.
2. Take a long branch, about six feet long, and use it as a ridgepole. Lay the ridgepole between the two forks.
3. Fill in the roof area with other straight sticks, with one end tied at the top and the other buried into the ground. You now have the skeleton of the shelter.
4. Cover the skeleton with whatever material is available: for example, spruce twigs, grass, bracken fern, and large leaves. Always start at the bottom of the shelter and work upwards when thatching so that if it rains, the water will run over the joints and will not leak through onto you. Try to make life as easy as possible by using any standing or fallen timber, or a wall, as one side of the shelter.
5. If heat is needed, build a fire in front of your shelter.

Scout Pits

There are unlimited variations of scout pits. Some can be constructed as hybrids with other shelters, and some partially above ground. Utilizing natural pits can significantly reduce the dig time. You can, for example, use a pit created where a tree has fallen over and pulled out the earth with its roots.

1. Find a flat, dry piece of land and dig a rectangular coffin shape that's a little wider and longer than your body. Preserve the top six inches of soil to serve as the ceiling.
2. Dig down to the length of your hips.
3. Dig a wider six-inch-deep ledge around the top to hold the roof in place.
4. Create a ceiling made of logs that at are a minimum of three inches in diameter. Leave a small entrance hole.
5. Cover the logs with soil and leaves.

≈ **Tarp lean-to**

≈ **Subterranean shelter covered with a tarp.**

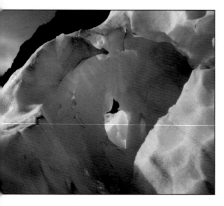

⌃ **Snow cave**

Ice and Snow Shelters

A snow cave, hut, or igloo can provide vital warmth and shelter in the snow.

Snow Cave

Find a wind swale, or cornice face, that allows you to dig horizontally. This will save you the time of digging down and then sideways.

If you find a good semivertical face, test to see if you can cut uniform blocks of snow with your shovel. If so, start with a fairly large opening, and enlarge into your sleeping chamber as soon as practical. When the cave is dug, wall the opening back up with the blocks of snow, leaving a small entranceway.

Be sure your sleeping chamber is higher than the door so that it can trap warm air from your body.

Before starting work, strip off all nonessential clothing so it stays dry, especially any down gear. If you have hardshell layers, wear them over nearly nothing while you're working. Once you're digging inside the entrance tunnel and beyond, you'll be amazed at how warm you become.

Once you are in the cave for the night, place your dry clothing next to your skin and your damp clothing away from your body. Use your backpack as a seat. Keep your boots on but loosen them as much as possible. If you feel your feet getting cold, take the liners out of the boots and store them under your clothing. DO NOT leave your liners in your boots, as you may be borderline hypothermic in the morning, and placing your cold feet in frozen boots could cause frostbite.

If you have a stove, you need a ceiling vent in your cave to prevent carbon monoxide poisoning. If you don't have a stove and your cave is large with an open doorway you probably don't need a vent, but consider a small one anyhow to keep the air drier and fresher. You can always plug it with a spare piece of clothing or with snow.

Improvised Snow Trench

If you're short on time and you have skis with you, simply dig a trench, lay your skis over it, and cover the skis with snow blocks. Keep the trench as small as possible so you can plug the end with a backpack or snow blocks once you're inside. The trench works better if you have a ground cover and sleeping bags, while the snow cave works better if you're short on gear. The main problem with building a snow cave is getting wet. If you're short on clothing, this is a major concern.

Note: In extreme cold, DO NOT use metal, such as an aircraft fuse-lage, for shelter. The metal will conduct away from the shelter what little heat you can generate.

Snow Shelter

1.   Clear a circular area in the snow about seven or eight feet across.
2.   Use a shovel or other digging tool to mix up the snow in the clearing, making sure to bring snow from bottom layers up higher and vice versa. Mixing snow of different temperatures will facilitate the hardening process, which is called sintering.
3.   Make a six-foot-high pile of snow on top of the clearing and shape it into a dome. The snow should be heaped, not packed.
4.   Allow the mound to sinter for one to three hours depending on the weather and snow composition.
5.   Begin to hollow out the mound once it has hardened. Dig straight in to create your initial opening; then dig at an upward angle in order to make an elevated sleeping area. This will allow cold air from inside to flow down and out of the shelter.
6.   Use the snow you dig out to make a windbreak in front of the entrance, or heap it onto the exterior of the shelter to thicken its walls and increase the available interior space.
7.   Smooth out the interior walls and ceiling when the hollowed area is large enough. The walls of your shelter should be at least one- to two-feet thick.
8.   Poke a ventilation hole through the top of the dome using a ski pole or long stick. Make sure this hole stays clear of ice and snow.
9.   Use a ski pole, sticks, or other large clearly visible item to mark the outside of the entrance in case it gets covered up while you're away. Keep your shovel inside while you sleep in case you need to dig your way out in the morning.
10. Use your pack to block the entrance of the shelter, but leave space for air to flow in and out.
11. Don't cook inside your shelter. This can cause a lethal buildup of carbon monoxide, even with a ventilation hole.
12. Building a shelter is hard work, so expect to sweat. This can cause hypothermia. If you have an extra set of dry clothes, change into them after you've finished building your shelter.

⩗ **Matches**

The Importance of Insulation

Knowing how to insulate yourself with layers of material can mean the difference between life and death. What you want to do is trap air between your body and outside. Fibrous plants, grasses, layers of bark, pine needles, leaves, wood, and even snow are types of materials that you can use to insulate yourself and your shelter.

To insulate your shelter, use the materials to build a thick layer over and inside your dwelling. Layering the materials will help trap air and keep more heat inside your shelter. Make sure you also use a thick layer of insulating materials inside your makeshift mattress. Doing so will insulate your body from the cold ground.

Urban Insulation

Urban environments contain numerous types of materials that you can use to keep warm—including cardboard, foam, cloths, plastic, bedding materials, and so on. In an emergency you can use these to turn a small room or closet into an insulted fortress. Couch cushions, blankets, towels, and mattresses can all be used to add extra insulation to your little area. Line your clothing with crumpled newspapers, paper towels, or any other insulating materials that you can find.

Fire Starting

Knowing how to build a fire is an essential survival skill.
Fire is good for:

1. Keeping warm
2. Boiling water
3. Drying wet clothes
4. Keeping insects and some animals away
5. To signal your position
6. Cooking

Always have at least two methods of starting a fire with you at all times.

Building a Fire

To build a typical campfire, you'll need three types of fuel: tinder, kindling, and logs. Always have twice as much of each as you think you'll need ready before you ever strike a match. The most difficult

A Survival, Evasion, Resistance and Escape (SERE) instructor with the Center for Security Forces builds a fire for the students to sit around during a fire building lesson at a training site in Warner Springs, Calif. ⩗

⌃ **Leaves**

⌃ **Logs and sticks**

⌃ **Logs**

part is getting the first flame to take to your tinder. Once you have a nice little pile of tinder material burning, it's relatively easy to get the rest of the fire going—first with kindling (big sticks), then with logs.

Collect dry wood. Start by looking for the dead branches at the very bottom of evergreen trees. Take the smallest branches and shred them with a knife or your fingers to use as tinder. Anything that will ignite quickly is a good source of tinder—dead grass, dried moss or fern, leaves, or a strip of cloth from the tail of your shirt.

Place your tinder in the center and then build a teepee of small dry twigs around it. Once this is burning, slowly feed your fire with larger and larger pieces of wood. Always make sure the fire is burning freely before you progress to a larger piece of wood. Once this fire is burning, do not let it go out. Aside from providing heat and protection, it can also act as a signal to anyone who is searching for you.

⌃ **Wood chips**

Tinder

Sources of good tinder include a multitude of mosses, grasses, and other thin and fibrous materials that can be easily ignited. They need to be dry. When walking along in the woods, collect wispy looking materials and put them in a shirt pocket, as body heat dries them out in a hurry. Here are some great sources of tinder that will light in just about any conditions:

- Cat-o'-nine tails. The large bulb at the top of this plant has enough "fluff" to start a LOT of fires.
- The large, black, lumpy growths on the sides of birch trees are caused by a type of fungus that burns very well. Each lump is orange to

⌃ **Grass**

151

≈ **Moss**

≈ **Cattails**

≈ **Chaga mushroom on birch tree**

≈ **Bush**

≈ **Shrub**

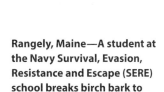

Rangely, Maine—A student at the Navy Survival, Evasion, Resistance and Escape (SERE) school breaks birch bark to start a fire. ≈

brownish on the inside and can be ignited with a spark to form a very nice coal. This material can also be used to transport fire from place to place.

- Low-lying, gnarly pine shrubs and trees (common in sandy soils) build up amazing amounts of sap. The wood becomes infused with it and is very flammable. Dead branches, in particular, are loaded with sap, and you should use these, as even a small piece can be used to start many fires. Shavings from this type of wood will ignite with nothing but a good spark. A little goes a long way.
- Pocket lint is very flammable. It only takes a spark.

There are several types of tinder that are easy to prepare and will serve you well if you find yourself trapped in the wilderness. Dryer lint and cotton balls, for example, both work well, especially when they're mixed with Vaseline. Heat the Vaseline (either in a microwave or in a pan on the stove) until it turns to liquid, then mash as much dryer lint or cotton balls as will fit and soak up the liquid. These can be kept in a plastic bag, aluminum foil, Altoids type container, or any other small container until ready to use.

Building a Fire in the Snow

This can be a little tricky, but you'll be successful if you follow these basic rules.

- Most fires will quickly heat the surrounding area, but when wind is present, most of the heat will be carried off. A fire in the wind is also going to consume about twice as much wood. Make sure you find a place to build your fire that's sheltered from the wind and elements.

- Gather all of your wood first and then organize it by size so you'll be able to find the right piece when you need it.

- Just because wood is buried under snow, that doesn't mean it's not dry enough to burn, especially if the snow is light and fluffy, which means it has less moisture content.

- Break a stick to see if it's dry inside. If it cracks, it most likely is. But if you're hiking after a winter rain, that crackling snap could be ice. If that's the case, you'll need to look for dry wood in protected areas, like under thick vegetation or in the hollow of an old tree stump.

- Sample wood from different places around your site. Keep track of what wood you found where, so you'll know where to return for more of the good stuff.

- Wet or damp wood can take a long time to get started. That's why you should always take some kind of fire starter with you. You'll find fire starters at most outdoor sports stores, army-navy stores, or at convenience stores in many rural areas. Look for tubes of fire ribbon, balls of wax mixed with sawdust, or tablets made of petroleum.

- You can also prepare your own tinder out of laundry lint or cotton balls as described above.

- Don't bother using toilet paper for tinder, as it burns for only a second.

- Pine needles and birch bark are great fire starters. Look for downed stumps.

- Stove fuel can give your fire the kick it needs to get going. Put the fuel on the fire before you light it, never after. Then toss in a match—and stand back!

- If the snow isn't too deep, dig a hole to make the fire on solid ground. If the ground is completely covered with very deep snow, tamp down the snow so it forms a slight depression with a solid, hard platform in the middle. Then put a layer of wood down on the snow, and build your fire on top of that. Otherwise, your fire will sink into the snow and go out before it gets going.

- When the fire is roaring, place damp wood around it so the heat from the fire will dry it out. Now you'll have a stash of dry wood for later.

⌃ **Rangely, Maine—A student at the Navy Survival, Evasion, Resistance and Escape (SERE) school builds a fire.**

⌃ **Collect fine twigs and then ignite the bundle off the ground.**

⌄ **Winter cooking fire**

Types of Fires

Lazy Man Fire

Maintaining a fire is just as important as starting one. If you're in a survival situation, you always want to save energy. Don't spend your time chopping firewood. Instead, feed large branches and logs into the fire, and let the fire do all the work. As the logs burn, move each farther into the fire. It's amazing how much wood you can gather when you're not wasting time chopping or sawing.

Teepee Fire

Build it with standing lengths of wood with tinder and kindling in the middle. The teepee fire provides a steady, hot heat source required for a reflecting oven. It requires a steady supply of medium-sized pieces of wood.

Pinwheel Fire

Lay one-to two-inch-diameter pieces of wood in a pinwheel pattern with tinder and kindling in the middle. This is an ideal fire for cooking with a fry pan. Build it inside a ring of rocks to hold your fry pan.

Log Cabin Fire

Stack four- to six-inch diameter pieces of wood in a crosshatch pattern. Provides lots of air circulation and results in quick supply of cooking coals for roasting or grilling meat.

Keyhole Fire

The keyhole is a great multipurpose fire when you have a larger group of survivors. Construct a rock fire pit in the shape of a keyhole. Next, build a teepee fire in the round part. At the end, build a log cabin fire. The tall flames of the teepee fire will provide light and heat once the coals of the log cabin fire die down.

Dakota Pit Fire

This is an efficient fire that uses very little fuel and can warm you and your food easily. Having it contained in a hole makes it is easy to hunch over for warmth or to place food or water over it for cooking. The second hole is to allow oxygen to get to the fire, thus preventing it from being easily smothered. The scale of the fire depends solely on the size of the pits you dig.

Note: This type of fire does not throw off much light and is primarily used for warmth.

Having selected a likely area in which to dig the fire hole, first remove a plug of soil and plant roots in the form of a circle about ten or twelve inches in diameter. Continue digging straight down to a depth of about one foot, being sure to save the plug and the soil you removed for replacement later on.

⋏ **Students in the Survival, Evasion, Resistance and Escape (SERE) course dig holes to make Dakota fire pits as part of the food preparation lesson at a training site in Warner Springs, Calif.**

1. Extend the base of the fire chamber outward a couple of inches in all directions so that it can accommodate longer pieces of firewood. This saves time and energy in breaking up firewood into suitable lengths, and also has the effect of allowing larger and therefore hotter fires.

2. Starting about a foot away from the edge of the fire pit, dig a six-inch diameter air tunnel at an angle so that it intersects with the base of the fire pit. The effect is a jug-shaped hole at the base of which you place firewood. The neck of the jug will serve as a chimney of sorts, the function of which is to increase the draft and concentrate the heat of the fire into the small opening.

3. Now it's time to make the fire hole airway. First determine the general direction of the wind, as you want to construct the airway on the side of the hole that faces the wind.

4. Dig your six-inch-diameter airway tunnel starting about one foot away from the edge of the fire hole. Angle its construction so that the tunnel intersects with the base of the fire chamber. Be sure to save the plug containing the vegetation and roots as well as the loose soil that you remove.

5. Partially fill the fire pit chamber with dry combustible kindling materials and light the fire. Gradually add sticks so that a strong hot fire is maintained.

⋎ **Pit fire**

10

FOOD AND HUNTING

"Luck favors the prepared mind."

—*Louis Pasteur*

Since food ranks third on the basic survival skills priorities list, don't worry about looking for food until you have a good water source and adequate shelter.

Although you can live three weeks or so without food, you'll need an adequate amount to stay healthy. Without food, your mental and physical capabilities will deteriorate and you'll become weak. Food replenishes the substances that your body burns and provides energy, vitamins, minerals, salts, and other elements essential to good health.

The average person needs approximately 2,000 calories per day to function at a minimum level. To varying degrees both plant and animals (including fish) will provide the calories, carbohydrates, fats, and proteins needed for normal daily body functions.

Most natural environments are filled with a variety of items that can meet our nutritional needs. Wild plants often provide the most readily available foods, though insects and small wild game can also support dietary needs in a survival situation.

Meat and fish are good sources of protein and fat and provide virtually everything a long-term survivor would need. However, at the first stage of a survival situation, edible plants are the most appropriate diet as plants are easily accessible and contain the necessary carbohydrates.

In a survival situation, take advantage of the food sources available. Try to vary your diet to make sure you get the appropriate proportions of fat, protein, carbohydrates, minerals, and vitamins.

Seek the more abundant and more easily obtained wildlife, such as insects, crustaceans, mollusks, fish, and reptiles, while you are preparing traps and snares for larger game.

Basic Food Survival Rules:

1. If it walks, crawls, swims, or flies, it is most likely safe to eat and will provide the nutrition and energy your body requires.
2. ALL fur-bearing mammals are safe to eat and will provide you with nutrients and calories.
3. ALL six-legged insects are safe to eat and will provide you with nutrients and calories.
4. DO NOT eat spiders.
5. Almost all freshwater fish and birds are safe to eat and will provide you with nutrients and calories.

6. Use EXTREME CAUTION with plants. Don't eat them unless you know they're safe.

7. DO NOT eat mushrooms, unless you are absolutely certain it is nontoxic, or any plant that has a milky sap.

Food Tips

A single emergency food bar can contain up to 3,600 calories and is designed to provide enough nutrition to last up to three days. Stash a few of these in your E&E kit, your second line gear, or in your go-bag.

Also easy to carry and useful are beef and chicken bouillon cubes. On a cold night out in the wild, a cup of hot broth will warm you up nicely.

Hard candy, i.e., Jolly Ranchers, offers a quick hit of sugar, which can be very helpful.

Edible Insects

Most insects are rich in protein and fat, the two most vital nutritional needs for survival. Ants, grubs, grasshoppers, dragonflies, worms, and centipedes are edible.

Some aren't the most appetizing and some taste pretty good. A good way to get over your natural resistance to eating insects is to dry them by the fire and add them to whatever you cook. I recommend that all insects be boiled or roasted to kill parasites.

Insects with bright colors should be avoided, as they might be toxic.

Any of the creatures listed below, once cooked, can be served with soy sauce or salt or mixed into a stir-fry or stew made of plants.

Grasshoppers

According to entomologists, a single large grasshopper is comprised of sixty percent protein and 6.1 grams of fat. Eating a handful of them roasted (not raw) is nearly the equivalent of consuming a hamburger. Crickets are second best. Remove the legs and wings, then roast them on a rock slab in the center of your fire for twenty minutes until crispy. Boiling for five minutes is another good way of cooking them. To catch them in the wild, use a three-foot section from a flexible, green willow shoot and swat them like you would a fly.

Ants

Boil the pupae (whitish eggs found in the nest) to make a hearty stew. The best way to collect pupae in large numbers is by carefully digging into the top layer of an anthill during the early morning. Make sure to avoid

⌃ **Hard candy**

⌃ **Flame skimmer dragonfly**

≈ **Grasshopper on a flower**

≈ **Ant**

≈ **Grubs**

≈ **Pine cones: About twenty species of pine trees grow pine cones that have seeds large enough to be food for humans. Place the pine cone next to a fire so it opens up. Then the seeds can be harvested easily and eaten.**

≋ **Puffball mushroom**

fire-ant mounds! One or two scrapes off a small section of the hill should expose the egg chamber. After collecting the eggs, cover the mound back up with dirt so the colony can recover.

Grubs, larva, worms

Earthworms can be dried like jerky and added to stew. Also, grubs found under or in rotten logs are relatively easy to collect, and can be added to a stew. Grubs, worms, and larva can also serve as excellent fishing bait.

Plants

Plants, roots, and green vegetables can provide carbohydrates and enough protein to keep the body functioning at normal efficiency, even in the arctic. Nuts and seeds are also an excellent source of proteins and natural oils.

Edible Plants

Be careful and don't eat any plant that you're not sure is safe. Some plants will poison you on contact, ingestion, or by absorption or inhalation. Also, many edible plants have deadly relatives and look-alikes.

As a general rule, before you head out into the wild, know at least three edible plants in the area that you are traveling. Learn them by their leaves, stalks/stems, and roots, as all three are part of the plant, might have use, and will confirm identity of the plants. Take into account the seasons when they grow and soils where they grow, and any relative plants that are around that might lead you to them.

I also recommend knowing at least three plants that are medicinal in some significant way and that are easily identified, readily available, and most useful based on your needs.

≈ **Nettle**

≈ **Acacia**

≈ **Agave**

Common edible plants include cattail roots, acorns, clover, dandelions, almost all grasses that are seed bearing, and the inner bark of trees such as poplar, willows, birches and conifers.

Common and Abundant Plant Food Sources in North America

A **manzanita bush** is a great food source in the wild. The Mono Indians used them for their fruit, which they dried or used for drinks. If you come across the bush, chew the outer bit of the fruit, then spit out the hard seeds. The fruit is high in vitamin C.

Cattails are known as the "supermarket of the swamp." No matter the season, there are always edible parts available on the cattail plant. The roots, shoots, and pollen heads can all be eaten.

The inner bark of **conifer trees**, known as the cambium layer, is full of sugars, starches, and calories. It can be eaten on most evergreen, cone-bearing trees except for yews, which are all poisonous and are identified by red berries). The inner bark should be scraped out and cooked to convert the fibers into a more digestible form.

All **grasses** are edible. The leaves can be chewed and the juices swallowed—though be sure to spit out any indigestible fibers.

All **acorns**—the nuts produced by **oak trees**—can be leached of their bitter tannic acids and eaten, providing an excellent source of protein, fats, and calories. To rid the acorns of tannins, place them in a net bag in a stream for a day, or put them into several changes of boiling water. White oaks have the least amount of tannins and therefore the best flavor.

Again, be sure to properly identify any plant you plan on consuming. Many plants can be difficult to identify and some edible plants have deadly poisonous look-alikes. If you cannot identify the plant, DO NOT eat it.

≈ **Amaranth**

≋ **Chicory**

⌃ **Acorns**

Edibility Test

The best way to determine what is and is not safe to eat in the wild is to develop a sound knowledge of the plant families that grow in the region you're traveling in.

Some parts of a plant may be safe to eat and other parts of the same plant may be poisonous. Be sure to treat different plant parts as separate entities. For example, common fruits such as apple, tomato, and mango are edible, while parts of the plants they grow on are toxic.

Plants growing in water may have *Giardia* on them. If you have a very sensitive digestion, you should avoid using wild plants as food.

Be especially cautious of any plant that has a bean pod or looks like a tomato, potato, or morning glory. And avoid any plant or plant part that smells like almonds or root beer, or which may have developed mold.

If you're stuck in the wild, are not familiar with the plants in the environment, and have no other available sources of food, follow this procedure for testing plant sources before you consume them.

1. Test only one part of a potential food plant at a time. Separate the plant into its basic components—leaves, stems, roots, buds, and flowers.
2. Smell the plant part for strong or acidic odors. Remember that smell alone does not indicate whether a plant is edible or inedible.
3. Do not eat for eight hours before starting the test.
4. During the eight hours you abstain from eating, test for contact poisoning by placing a piece of the plant part you are testing on the inside of your elbow or wrist. If your body does have a reaction, it will usually occur within fifteen minutes.
5. During the test period, take nothing by mouth except purified water and the plant part you are testing.

⌄ **Dandelions**

⌄ **Poplar trees**

↟ **Fir trees**

↟ **Willow tree**

↟ **Birch tree bark**

6. Select a small portion of a single part and prepare it the way you plan to eat it.

7. Before placing the prepared plant part in your mouth, touch a small portion to the outer surface of your lip to test for burning or itching.

8. If there's no reaction on your lip after three minutes, place the plant part on your tongue, holding it there for fifteen minutes.

9. If there is no reaction after fifteen minutes, thoroughly chew a pinch and hold it in your mouth for fifteen minutes. Do not swallow.

10. If no burning, itching, numbing, stinging, or other irritation occurs during the fifteen minutes it's in your mouth, swallow the food.

11. Wait for eight hours. If any ill effects occur during this period, induce vomiting and drink a lot of water.

12. If no ill effects occur, eat a quarter of a cup of the same plant part prepared the same way. Wait another eight hours. If no ill effects occur, you can consider the plant part as prepared safe to eat.

Make sure you have other options.

Flood, fire, and droughts could all pose a serious risk to plant life. You need to have a backup plan for those times when eating plants may become impossible. Make sure you have a good emergency food supply and a way to fish, hunt, and trap game.

Animal Foods

Meat contains more protein than plant food and might even be more readily available in some places. But to acquire meat, you need to know the habits of, and how to capture, the various wildlife.

⌃ **Aleutian cackling goose**

⌃ **A male and female pair of Northern bobwhite quail**

⌃ **Seagull**

⌃ **Turkey toms**

Birds

All birds are edible. Game birds such as grouse and pheasants can be captured using snares or hunting implements such as a throwing stick, though it can be difficult if you're not familiar with the proper techniques.

Bird Eggs

In the spring, bird eggs are an excellent source of survival food. Not only are they high in nutritional value, they're also convenient and safe. Eggs can be boiled, baked, fried, or, if need be, raw. The obvious place to look for them is in a bird nest. However, not all birds build a nest. Some lay their eggs directly on the ground or in a hole.

Small Mammals

Small mammals including squirrels, rabbits and mice can be captured with practice. Traps and snares are often most effective, though a throwing stick can be used too.

⌄ **Jackrabbit**

⌄ **Mouse**

≈ **Bats**   ≈ **Prairie dogs**   ≈ **Rabbit**   ≈ **Squirrel**   ≈ **Raccoon**

Hunting and Gathering

Hunting for food in the wilderness is an important and often misunderstood skill that was second nature to our ancestors, but has been largely lost by modern man as he's evolved. The good news is that all of us are born with the senses and abilities to be good hunters.

Trapping and wild game procurement are not skills that can be acquired in the comforts of your backyard. They come with trial and error and plenty of practice out in the wilderness.

Those who take the time to learn and practice hunting and gathering skills will come to realize that wilderness landscapes that once seemed inhospitable are in reality generous lands that will sustain those who know where and, most important, how to look.

If you want to be able to feed yourself reasonably well in the backcountry, focus on developing the following basic skills:

1. Proficiency with a firearm
2. Fishing methods such as angling and using cast nets and trotlines
3. Knowledge of common edible plants in your region of travel
4. How to use traps and snares

≈ **Fish with worm**

≈ **Carp in net**

≋ **Gaur**

Living Off the Land

The initial emphasis for a survivor should be on small game and not big game animals such as elk, moose, and deer. On any given day in the wild, you are going to come across a greater concentration of rabbits, squirrels, woodchucks, marmots, raccoons, and other smaller critters than you will big game. For the survivor, these animals will provide sustenance until you can procure larger game.

≈ **Macaque**

≈ **Red muntjac deer**

≈ **Sambar**

Trapping

Trapping is time-intensive in the beginning, but the payoff is well worth it if you know what you're doing. Once learned, it is one of the most efficient means of feeding yourself in the wild.

Trapping is a numbers game. The more traps you set, the better off you will be.

Set out as many traps as you are physically capable of. There's no such thing, under survival conditions, as overdoing it. If you set up a primitive trap line for survival purposes, make sure it contains anywhere from ten to twenty or more traps (snares, deadfalls, and so on.).

Deadfalls and Snares

Snaring and deadfall traps are both excellent means of procuring meat—whether small game or large animals such as deer. Deadfalls work by crushing your intended prey while they attempt to eat the bait that you have placed on the trap mechanism. They are very effective for squirrels, chipmunks, and pack rats. In ancient Europe, hunters even used large deadfalls to bring down wolves and bears!

There are four things to consider when setting a deadfall trap:

1. Carving a precision trap and then finding the proper rock or log.
2. Knowledge of both the behavior and tracks of your intended prey.
3. Setting the trap in the correct area for your intended prey.
4. Using a bait or lure that will appeal to as many of the animal's senses as possible.

Primitive deadfalls for small game require willow, tamarisk, or other straight stalks. You can even use juniper logs that are split out into straight timber. The best way to carve deadfalls is to cut three dozen willow shoots and then spend the evening whittling twenty or more traps. Don't worry about placement or rigging these up in the field. Just sit and practice carving so you ingrain the moves in your head and hands.

The next step is to practice setting up your traps under nonsurvival conditions. You want to create a stable trap with a figure-4 trigger that will cause the trap to collapse, even to the slightest touch. As a weight, use a heavy log or rock, which should be two to three times the body weight of the animal you intend to trap.

How to Make a Figure-4 Trigger:

1. Rest a rock, log, or other weight on the end of a diagonal stick. The weight will supply downward pressure to the end of the stick.
2. The diagonal stick should rest and pivot on the vertical stick. This will keep the diagonal stick from slipping to the left.
3. The diagonal stick is also held in place by having the end rest in a notch.
4. The pressure of the diagonal stick will pull the horizontal stick toward the left.
5. The horizontal stick is held in place by the notches, which is where the actual trigger mechanism is located.
6. Please your bait at the end of the horizontal stick so it is directly under the weight.

≈ **Widget deadfall. This is a variation on the figure-4.**

When the animal takes the bait at the end of the horizontal stick, the trigger will release. This will cause the horizontal stick to fall to the ground, and the diagonal stick to flip up and out in a counterclockwise arc so that the weight will come crashing down on the animal, killing it instantly.

Make sure the vertical stick rests on a hard surface, such as a flat stone. Otherwise it may dig into the ground and won't fall out of the way when the trap is triggered. I also recommend putting something hard on the ground under the whole trap or placing the trap on rock. Otherwise, the deadfall, when it falls, may only injure the animal by pressing it into the soft ground.

Paiute Deadfall

The Paiute deadfall is the fastest deadfall you will find, and used throughout the American Southwest and Africa for catching or killing small game. If you use this type of trap with a box or crate instead of a deadfall, you can trap animals rather than kill them.

Unlike the figure-4 trigger, the Paiute deadfall uses cordage.

Materials required:

- A small length of cord.
- Two sticks equal in diameter, one roughly eight inches long, the other roughly 6 ½ inches long.

- A long skinny trigger stick (you can trim it to fit).
- A three-inch toggle stick (which should be a little thicker than your trigger stick).
- A deadfall (rock or heavy log) or a box or crate.

1. Attach the cord to one end of the 6 ½-inch-long support stick, and the other end of the cord to the middle of the three-inch toggle stick.

2. Find a place to set up your deadfall trap, one where you think your prey is likely to be found. Make sure the trap is set up on hard ground. Push your eight-inch-long upright stick about two inches into the ground and lean it slightly in the direction of the deadfall.

3. Set your trap with bait. Since you want your prey to pull down on the trigger stick, either rub your bait into the trigger stick or attach it to a piece of string that is tied to the trigger stick. Take your support stick and place the end without the cord on top of the upright stick. Then place the edge of your deadfall on the other end of the support stick. Hold the cord tight so that the support stick doesn't let the deadfall down. Then bring the cord around the upright and hold it there.

4. Placing the trigger stick can be frustrating. Be patient. Place one end of the trigger stick against the toggle so that the toggle cannot unwrap from around the upright. Now take the other end of the trigger stick and fit it into the underside of the deadfall. If you need to, trim the trigger stick so that it fits perfectly between the toggle and deadfall. Now you're ready to catch game.

Survival Snares

Snares are the simplest means of procuring small animals like rabbits and squirrels, and can also be used to catch larger game. A snare consists of a loop (noose) of cord or wire placed across an animal trail, a trigger mechanism, and a spring stick of some sort that will quickly jerk the animal up into the air when triggered.

There are two basic types of snares. The first (holding snare) holds the animal at ground level and may or may not strangle the victim. The second (flip snare) design will flip the animal into the air and cause it to be strangled. While both are easy to make, each design has strengths and weaknesses.

Both designs require a loop made of wire, cord, string, or vine to tighten and hold the animal. The loop should have some freedom of movement to allow it to tighten as the animal struggles or moves forward into the snare.

With both types of snares, make sure to set the loop diameter for the type of animal you hope to catch.

- For rabbit, the loop should be about four inches in diameter and placed about two inches above the trail.
- For squirrels, the loop should be about three inches in diameter and two to three inches above the trail.
- For beavers, make the loop about five inches in diameter and place it about one to two inches off the ground.

When making a holding snare, you have to secure the end of the snare wire (the end opposite the loop) to a bush, stake, or other stationary object. Make sure the snare is secure, and try to use brush, logs or other debris to funnel the animal toward your snare. The animal's head will enter the loop and, as it continues to move forward, the loop will slide and tighten until the animal can't escape.

With a flip-up snare, you also want to funnel animals toward your trap. The difference in this case it that when the animal's head enters the snare, it will eventually pull the wire far enough to trigger the flip-up part of the trap. At that point the animal will be flipped into the air and strangled. The diameter of the loop and the distances off the ground remain the same in this snare as in the other.

To make a trip snare, you need a flexible limb or bush, snare wire, a trigger, and a method to hold the trigger. I recommend a supple sapling. When bent over and secured with a triggering device, it will provide power to your snare. Select a hardwood sapling along the trail, and remove all branches and foliage if you can (as this will allow it spring faster). Use two forked sticks, each with a long and short leg. Bend the sapling or branch and mark the place where it meets the trail. Drive the long leg of one forked stick firmly into the ground at that point. Make sure the cut on the short leg of this stick is parallel to the ground. Tie the long leg of the remaining forked stick to a piece of cord secured to the sapling or branch. Cut the short leg so that it catches on the short leg of the other forked stick. Extend a noose over the trail. Set the trap by bending the sapling or branch and engaging the short legs of the forked sticks. When an animal catches its head

⌃ **A flip-up snare that yanks the critter off the ground and enables a quick, humane kill.**

in the noose, it pulls the forked sticks apart, allowing the twitch-up to spring up and hang the prey.

Baiting Your Traps

If you are out in the wilderness without bait, then use something exotic and uncommon to the area. If you are trying to trap a rabbit, don't use the raspberries off the bush near its burrow. Choose some succulent cattail roots from a nearby swamp. This will spark much greater interest than the greens it is surrounded by on a daily basis. Look inside your pack—any bagels or apples? The key here is to use something "exotic" from outside the animal's home region.

Prehistoric Trap Systems of Northern Arizona —by Tony Nester

The ability to procure food in a demanding landscape like the desert was possible for those who knew where and how to look and had a wealth of hunting and trapping skills. I believe trapping played a significant role in the menu of hunter-gatherers in the desert, though archeological site reports tend to focus on larger fauna and hunting implements such as the atlatl and bow.

Despite ranking low in the overall archeological interpretation of the Southwest, deadfalls and snares have been found in quantity at Basketmaker sites in Arizona and Utah and were used (and still are in some regions) at places like Hopi and Supai.

As experienced primitive trappers know, the use of deadfalls and snares is a very calorie-efficient method for obtaining wild game from the landscape.

Bird Snares

"We little boys made snares of horse-hair to catch birds. I learned to catch bluebirds with a hair from a horse's tail set as a snare on the upper stem of a sunflower stalk, with a worm for bait," said Hopi elder Don C. Talayesva in his biography *Sun Chief.*

I made a trap using a branch of willow with several (human and horse hair) snares attached. This trap was placed on the ground where invasive Eurasian doves land frequently to feast upon wild sunflower seeds.

Upon initial testing, I set the trap down in the evening and secured it with a line of yucca attached to a grapefruit-sized rock to prevent any airborne theft of the trap by a particularly pugnacious dove. At sunrise, my dogs informed me that there was a visitor in the trap. A groggy stumble outside revealed a large pigeon with one foot in the snare and the other foot attempting a wave to his freedom-loving pals above. The trap had worked and the pigeon was released unharmed by snipping the cordage near his foot.

Given the snare's low-tech nature and ease of construction, setting a few of these in heavily trafficked areas that birds frequent would provide a hunter with an easy meal with a minimum of energy expenditure. If one doesn't have access to birdseed (wild or otherwise) for bait, then placing the snares alongside water holes, where bird tracks have been found, would be the next best setup.

Another invention that doesn't show up in the ethnographic literature on the Hopi very often is an upright bird snare perched on a sunflower stalk. The setup is a miniature version of an Ojibwa bird snare from Canada and is used mainly for catching small birds such as bluebirds and sparrows, for their feathers and not their meat.

In DuPont Cave in Utah, a cache of 137 bird snares was found, while fifty-five bird snares were uncovered in a cave in Adugegi Canyon in Arizona; both were Basketmaker sites.

In my research I found that "snares consist of a stick measuring 50 to 60 cm. long by 0.5 to 0.75 cm. in diameter to which lengths of human hair or vegetable cordage have been secured at the distal end of each length of cordage is a small slip noose. Variation is restricted to the number of cords (from one to six) attached to the snare sticks." In my own fieldwork, bird snares that I have had success with each had five to eight cords attached.

In the western Grand Canyon, on Hualapai tribal land, there was an even simpler design in which "wild pigeons were caught in a snare, onu'k. This was a running noose of yucca twine. It was tied to a bush and would tighten up on the bird's leg" (from *Walapai Ethnography* by A.L. Kroeber).

Paiute Deadfall

"The Walapai have a deadfall trap, kweo'ne, in which they catch rabbits, rats, mice, and even snakes and lizards." The Paiute deadfall was used by the Supai in the Grand Canyon.

Traps are placed on the rodent trails in the fields to catch the squirrels and rats. Snakes and birds are sometimes found in these devices, but no attempt is made to trap or snare large game. It is baited with dried peaches or mescal pulp tied firmly to the trigger.

"The old people showed us how to make deadfalls to catch kangaroo rats, prairie dogs, porcupines, badgers, chipmunks, squirrels, and turtledoves. The men used heavy rock deadfalls for trapping coyotes, foxes, wildcats, and other large animals" (Hopi elder Don C. Talayesva in his biography *Sun Chief*).

≈ **Lizard**

Promontory Peg Deadfall

In Danger Cave, located two miles east of Wendover, Utah, archeologists uncovered more than sixty Promontory Peg components made from willow, milkwort, rabbitbrush, and other materials. One specimen even had a slice of prickly pear still impaled on the bait stick!

The Danger Cave specimens exhibited signs of spiral cuts on the platforms which might, as one researcher noted, have been to increase surface friction or create "threads" to allow for a better union between the two pieces.

On a twelve-day desert survival course I taught for the military, students made two Paiutes and two Promontory Peg deadfalls each. These were baited and placed among the rocky ledges in a canyon not far from our camp. The trap design that was initially successful, and which scored pack rats, was the Promontory Peg. After refinement (and more in-field application) during the following week, students experienced a more balanced success rate between the two trap systems. This was largely due, I believe, to continual practice with the delicate Paiute trigger system as well as more field experience with reading animal signs, which ensured better trap placement.

The Great Pack Rat Roundup

Every fall, during our annual five-week program in traditional skills, we spend a great deal of time on the area of food procurement, particularly teaching a variety of primitive traps. This season, we had the opportunity to test out a large-scale primitive trapline (in this case aimed at rodents). The region consists of forty acres of high-desert, pinyon-juniper at an elevation of 6,000 feet.

Three hogans and numerous wickiups on site had been infested with pack rats and we decided to have students set individual traps consisting of two Paiutes and two Promontory Peg deadfalls per person. Traps were baited with local flora such as currants, prickly pear fruit, and wild sunflowers. All traps were fenced in with twigs of juniper, save a small entrance. We agreed that, assuming the animals caught were healthy, we would utilize the meat in our stews and jerk any surplus. An adult pack rat doesn't weigh a lot, so there wasn't exactly enough for a juicy rat-burger.

There were six participants and we spread our traps over an area of approximately twenty acres. Care was made not to disturb any nests or droppings and everyone got a lecture on the dangers of zoonotic diseases such as hantavirus from deer mice droppings and bubonic plague from flea bites.

We determined that there were roughly six pack rats per acre, or approximately 120 rats over twenty acres. Add in the other small critters, such as rock squirrels, prairie dogs, cottontails, and jackrabbits, and a prehistoric trapper would be able to fill his stew pot using traps.

Traps were set in the afternoon and checked once before everyone went to sleep and once upon rising in the morning. The students had been practicing their trap skills for the past week, since the program began, and were already familiar with how to fine-tune their deadfall triggers.

Here are the results based on twenty-four deadfalls (two of each type) set out during a twelve-hour period (overnight):

Traps Set: 24
Traps Sprung (but without game): 7
Traps Sprung (with game): 4
Traps Unsprung: 13

Of the traps that were unsprung, some had bait that was missing while others were untouched. The traps that were sprung but without game (no kill) had the bait missing; several bait sticks were completely missing from the scene. The traps that were sprung with game all had flattened rats. Two of the latter traps were set inside the hogans that had the most rodent traffic. The remaining two traps that had game were set outside: one near the entrance of a large hogan and the other adjacent to a rock pile near an old wickiup. Regarding which traps were more successful in this very brief experiment, the Paiute deadfall accounted for three rats and the Promontory Peg accounted for one rat. The bait that was favored was fresh prickly pear fruit that had been gathered previously from outside the area.

After students collected rats from their traps, they immediately tossed each rat on the coals of the fire to burn off the fleas. We then divided up the rats and had one of these skinned and boiled up in a stew pot while the other two were tossed back on the coals to bake the meat in its skin, Hopi-style. After removing the latter group of rats, the crunchy, blackened carcasses were gutted to remove the innards and the meat consumed by picking past the skin. One rat was dried, skeleton and all, to show how to make jerky from small critters.

Again, be warned that pack rats and other rodents carry fleas and should be treated by immediately singing off the hair in the fire or by immersing the carcass underwater in a stream for thirty minutes.

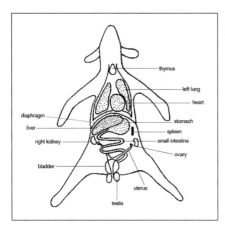

⋏ **Mammal anatomy diagram**

⋎ **Skinning a rodent**

Skinning and Gutting Your Kill

A "gut hook" or "belly zipper" is the curved hook that's located on the back of many hunting knives. Use this to hook under the skin and "unzip" the pelt without snagging on internal organs. If you don't have one on your knife, be very careful when opening any animal so as to not cut into the intestines or stomach and contaminate the meat.

You should have a basic understanding of the general location of a mammal's internal organs. Since all internal organs are connected at the throat and anus, by cutting at those two points you should be able to remove all of them in one big mass.

- **Esophagus**—Connects the mouth and the stomach. Because it contains a lot of cartilage and gristle, it's not really worth eating.
- **Stomach**—Contains harsh acids along with partly digested food. Unless properly prepared, don't eat it.
- **Lungs**—They're edible, but most people find them hard to eat.
- **Heart**—Solid muscle and good to eat if tenderized or boiled first.
- **Kidneys**—Good to eat.
- **Small intestine**—Contains a lot of harsh acids and partly digested food. Don't eat it unless properly cleaned, prepared, and cooked.
- **Liver**—Packed with vitamins and minerals and good to eat unless there are any discolorations or spots on it. If you find spots or discolorations, discard.
- **Large intestine**—Don't eat it.

Possible methods for catching a nonvenomous snake:

- Distract the snake with a stick. Grasp its tail and lift the snake upwards, leaving the front part of its body on the ground.
- Lay a large container on its side. Sweep snake in with a branch or a stick.
- Use a large bushy stick. Allow the snake to "hide" in your makeshift "bush." Tease the snake with another stick or by wiggling the bushy stick. The snake will often curl around the branches and twigs. Move your stick to where you want the snake to go and leave it there. The snake will leave when it feels safe.

- Look for an area that has shade and bugs. When you see the snake, slowly and very, very carefully, pick it up by the head. Bring it over to the bag or container.
- Grab the snake by the tail if possible, keeping the rest of your body as far away as you can. Pin the snake down by placing the stick directly behind its head and applying pressure to keep it from moving its head. Drop the tail and grab the pinned snake by the neck as close to the head as possible. Use your other hand to support the snake's body and keep it from thrashing about. A large snake can easily free itself from your grip if you let it thrash.

⩘ **Black snake**

Prepare a Snake for Eating

Snakes can be quite tasty and are a staple of protein, much like duck, pheasants, turkeys, and rabbits.

Instructions:

1. Remove the snake's head and neck up to three inches, clean with knife;
2. Slit the snake lengthwise up the middle and remove the insides.
3. Rinse and clean the inner part of the snake.
4. Remove the skin. Start where the neck was and slowly peel back.
5. Remove the tenderloins of the reptile. You do this by laying the snake belly down (like you would fillet a fish) and cutting the meat away from the back and ribs.
6. Place the meat in a bowl of fresh water with about one tablespoon of salt. Let this sit until you are ready to grill or fry.

Survival Fishing

Fish are a valuable food source, and all freshwater fish in North America are edible. In a survival situation, there a several ways to catch fish, including using a sharpened stick as a fish spear, or making a fish net for small minnows out of a T-shirt.

⩘ **Carp**

Angling

Angling is the method of catching fish that people are familiar with. You'll need a hook, line, rod, and a small weight to take the hook down. You'll also need a float to keep the bait off the lake or pond bottom. You can use any small floating object as a bobber, including a small piece of bark. You don't need a float in streams and flowing water.

Large dip net made from a hefty oak branch and the inner strands of 550 cord.

You should have a line and hook as part of your survival kit. Hooks can also be improvised from other kinds of materials such as safety pins, thorns, bones, and wood. In order to be successful at fishing, you need to know something about the behavior of the fish you're after. In general, the best time to catch fish is just before dawn, just after dusk, or when bad weather is imminent.

Like most all other living creatures, fish choose those places where they are most comfortable and where they can most easily find their prey. If it's hot and the water is low, deep shaded water is probably where you should look. In cooler weather, you probably will find your catch in a shallow place where the sun warms the water. Fish like to shelter under banks and below rocks or deadfalls.

Fish are more likely to take bait native to their water, so try to figure out what they eat. For example, insects and worms can be used as bait in practically any water. If one type of bait isn't successful, change to another.

Survival Fishing with a Spear

Spear fishing takes time, patience, and practice. For best results, use a forked spear and look for shallow water where your catch is visible. Slowly move the spear as close as possible to your catch, and spear it quickly.

Fishing Nets

In a survival situation, an improvised net can prove extremely effective. For instance, use an undershirt and a y-shaped branch to make a landing net. T-shirts and other pieces of fabric can also be used as nets to catch small fish.

Survival Cooking

Cooking is an important skill for those who spend time in the wilderness. Cooking not only makes many foods more appetizing to taste, but also ensures that parasites and bacteria are killed. The last thing you want is to get sick from food poisoning.

While it's certainly convenient to carry a compact camping stove in the wilderness, in many situations a cooking fire is more practical and allows a wider variety of cooking opportunities.

Hot Stone Cooking

This cooking method is ideal for fish, thin slices of meat, and eggs. Simply light a fire above a bed of nonporous stones. Don't use soft, porous stones with high moisture content because they might explode on heating.

High-altitude cooking

Let the fire burn for an hour or more, then brush away fire and embers with a stick or a handful of long grass. Cook your food directly on the rocks, the way you would using a frying pan.

Hot Stone Pit

This alternative acts as a primitive oven.

Dig a pit about two feet deep and two feet in diameter. Pack the pit bottom and walls and cover the bottom with hot fist-sized stones. Then add a thin layer of soil. Wrap your meat in fresh green plant leaves or moss. Add another thin layer of soil and more hot stones. Cover with earth or sand.

⌃ **Fish fillets skewered with green willow sticks bake near the coals.**

A Modern Hunter-Gatherer's Outdoor Calendar from Northern Arizona—by Tony Nester

In Arizona there are many life zones and a range of environments, from low desert at 1,200 feet to alpine at 10,000 feet. With such a range of elevations and life zones, there are unique opportunities to forage and hunt.

Each region of the country presents its own challenges and optimal times for harvesting, so you will have to look into what is available where you live and travel and create your own calendar. Below is an example of my personal calendar for hunting and foraging in northern and central Arizona by season.

Spring

Collect young cattail and bulrush shoots from riverbanks and riparian areas
Gather young amaranth, dandelion, curly dock, and goosefoot leaves for salads
Collect wild onions before flowering
Fish for trout and panfish
Hunt rabbits and raccoons

Summer

Gather currant berries and raspberries
Collect cattail pollen for making bread
Harvest purslane for salads
Collect medicinal plant leaves for drying
Collect crayfish
Fish for catfish and panfish
Hunt rabbits

Gather puffball mushrooms

Pick banana yucca fruits

Fall

Gather large quantities of gambel oak acorns and piñon pine nuts

Bow hunt deer and any other large game in the area

Hunt squirrels, rabbits, and raccoons

Collect medicinal plant roots for drying

Gather mesquite pods for grinding into flour

Harvest apples from local orchards

Winter

Hunt rabbits

Bow hunt deer (during late archery season)

Harvest willow for teaching trap-making and making fish baskets

Process dried acorns (from the fall) into flour

Refine skills, do research on mammals/trapping/plants, and wait for spring

As you can see, summer and fall offer a bounty of resources, especially in terms of nuts, acorns, and plant resources compared with winter and spring. I would encourage you to formulate your own region-specific calendar.

As your skills progress and you become more proficient at hunting and gathering, you will realize that what seems, at first glance, to be an inhospitable landscape is in reality a generous land that will sustain those who know where and, most important, how to look.

11

WEATHER

"Everybody talks about the weather, but no one
does anything about it."

—*Charles Dudley Warner*

Whether we realize it or not, most of us subconsciously "forecast" the weather. Say you look outside and see dark clouds approaching. Chances are you'll grab your rain gear as you walk out the door. If you're out on a trail run and are suddenly hit by a gust of wind, you're likely to glance up at the sky to look for other signs of ominous weather.

There are better ways to forecast the weather than relying on "subconscious forecasting" alone. In the mountains, desert, jungle, or at sea, an understanding of weather patterns is critical. The ability to accurately forecast and plan for changes in the weather could save your life.

Mountain Weather

Mountain weather can be extremely erratic and harder to read than in other areas. Conditions vary with altitude, latitude, and exposure to atmospheric winds and air masses.

The weather there can vary from stormy winds to calm, and from extreme cold to heat within a short time or with just a minor shift in locality. The severity and variance of the weather can have a major impact on your survival in the mountains.

Considerations for Planning

Mountain weather can work in your favor or become a dangerous obstacle to survival, depending on how well you understand it and to what extent you take advantage of its unique characteristics.

The clouds that often cover the tops of mountains and the fogs that cover valleys are excellent means of concealing movements that normally are made during darkness or in smoke. Tactically, limited visibility can be used to your advantage. Darkness can be your friend.

The safety or danger of almost all high mountain regions, especially in winter, depends upon a change of just a few degrees of temperature above or below the freezing point. Ease and speed of travel depend mainly on the weather. Terrain that can be crossed swiftly and safely one day may become impassable or highly dangerous the next due to snowfall, rainfall, or a rise in temperature. The reverse can happen just as quickly. The prevalence of avalanches depends on terrain, snow conditions, and weather factors.

Some mountains, such as those found in desert regions, are dry and barren with temperatures ranging from extreme heat in summer to extreme cold in winter. In tropical regions, lush jungles with heavy seasonal rains and little temperature variation often cover mountains. High rocky crags with glaciated peaks can be found in mountain ranges at most latitudes along the western portion of the Americas and Asia.

Severe weather can dramatically lower morale and intensify basic survival issues. Problems can be minimized if you've been trained to accept the weather and are properly equipped.

Mountain Air

High mountain air is typically dry, especially in winter, when cold air has a reduced capacity to hold water vapor. Because of this increased dryness, equipment does not rust as quickly and organic material decomposes slowly. The dry air also requires that you increase your consumption of water. That's because reduced water vapor in the air causes an increase in evaporation of moisture from the skin as well as a loss of water through transpiration in the respiratory system. Cold air tends to subconsciously discourage people from drinking enough liquids; always keep this in mind, and make a conscious effort to increase your fluid intake.

Pressure is low in mountainous areas because of the altitude. The barometer usually drops 2 ½ centimeters for every thousand feet gained in elevation (three percent).

The sun's rays are absorbed or reflected in part by the molecular content of the atmosphere. Rays are reflected at a greater rate at lower altitudes. At higher altitudes, the thinner, drier air has a reduced molecular content and, consequently, doesn't filter the sun's rays nearly as much.

As such, the intensity of both visible and ultraviolet rays is greater with increased altitude. These conditions increase the chance of sunburn, especially when combined with a snow cover that reflects the rays upward.

Weather Characteristics

The earth is surrounded by an atmosphere that's divided into layers. The world's weather systems occur in the lowest of these layers, the "troposphere," which reaches as high as 40,000 feet. Weather is produced by several factors: the atmosphere, oceans, land masses, unequal heating and cooling from the sun, and the earth's rotation. The weather found in any one location depends on factors such as the air temperature, humidity (moisture content), air pressure (barometric pressure), how that air is being moved, and if it is being lifted or not.

Air pressure is defined as the "weight" of the atmosphere at any given place. The higher the pressure, the better the weather will be. Conversely,

⌄ Barometer

the lower the air pressure, the more likely the weather will be problematic. The average air pressure at sea level is 29.92 inches of mercury (hg) or 1,013 millibars (mb). The higher the altitude, the lower the pressure.

High Pressure: The characteristics of a high-pressure area are as follow:

- The airflow is clockwise and out, otherwise known as an "anticyclone."
- Is associated with clear skies.
- Generally the winds will be mild.
- Depicted as a blue "H" on most weather maps.

Low Pressure: The characteristics of a low-pressure area are as follow:

- The air flows counterclockwise and in, otherwise known as a "cyclone."
- Is associated with bad weather.
- Depicted as a red "L" on most weather maps.

Air from a high-pressure area is trying to flow out and equalize its pressure with the surrounding air. Low pressure, on the other hand, builds up vertically by pulling air in from outside itself, which causes atmospheric instability, resulting in bad weather.

On a weather map, these differences in pressure are depicted as isobars, which resemble contour lines and are measured in either millibars or inches of mercury. Areas of high pressure are called "ridges" and lows are called "troughs."

Wind

The ridges and passes found in high mountains are seldom calm. Inversely, strong winds in protected valleys are rare. Normally, wind speed increases with altitude since the earth's frictional drag is strongest near the ground. This effect is intensified by mountainous terrain because winds accelerate when they converge through mountain passes and canyons. This funneling effect can cause the wind to blast with great intensity on an exposed mountainside or summit. Usually, the local wind direction is controlled by topography.

The force of wind quadruples each time wind speed doubles. So a wind that blows at forty knots pushes four times harder than a wind blowing at twenty knots. As wind strength increases, gusts become more important and may be fifty percent higher than the average wind speed. When wind strength increases to a hurricane force of sixty-four knots or more, you should lay on the ground during gusts to avoid injury and continue moving during lulls. If

« **Weather map**

Weather Forecast for Monday, February 02, 2004
DOC/NOAA/NWS/NCEP/Hydrometeorological Prediction Center
Prepared by Hatchett/Eckert based on HPC, SPC, and TPC forecasts.

a hurricane-force wind blows where there is sand or snow, dense clouds will fill the air. Rocky debris or chunks of snow crust will be hurled near the surface. In winter, or at high altitudes, you need to be constantly aware of the wind-chill factor and associated cold-weather injuries (see chapter 4).

Winds are the result of the uneven heating of the air by the sun and rotation of the earth. Much of the world's weather depends on a system of winds that blow in a set direction.

Above hot surfaces, air expands and moves to colder areas where it cools, becomes denser, and sinks to the earth's surface. The results are a circulation of air from the poles along the surface of the earth to the equator, where air rises and moves to the poles again.

This heating and cooling dynamic, coupled with the rotation of the earth, causes surface winds. In the Northern Hemisphere, the three prevailing winds are:

1. Polar Easterlies—winds from the polar region moving from the east. This is air that has cooled and settled at the poles.
2. Prevailing Westerlies—winds that originate from the west. This is where prematurely cooled air, due to the earth's rotation, settles to the surface.
3. Northeast Tradewinds—winds that originate from the north and/or from the northeast.

The jet stream is a long, meandering current of high-speed wind often exceeding 250 knots that flows in the transition zone between the troposphere and the stratosphere—known as the tropopause. These winds generally blow from a westerly direction, and dip down and pick up air masses from the tropical regions, then move north and pull air down from the polar regions.

Air, which makes up wind and moves in parcels is called "air masses." These air masses can vary from the size of a small town to as large as a country. They're named according to their area of origin:

1. Maritime—over water.
2. Continental—over land.
3. Polar—north of sixty degrees north latitude.
4. Tropical—south of sixty degrees north latitude.

When you combine these parcels of air with their descriptions, you get the four types of air masses:

1. Continental Polar—a cold, dry air mass.
2. Maritime Polar—a cold, wet air mass.
3. Maritime Tropical—a warm, wet air mass.
4. Continental Tropical—a warm, dry air mass.

Two types of winds are found in mountain environments.

1. Anabatic Winds (valley winds)—These are winds that blow up mountain valleys to replace rising warm air. They're usually light.
2. Katabatic Winds (mountain winds)—Winds that blow down mountain valley slopes caused by the cooling of air. They can be strong.

Humidity

Humidity is defined as the amount of moisture in the air. All air holds water vapor even when it can't be seen. But air can hold only a limited amount of vapor. The warmer the air, the more moisture it can hold. When a body of air reaches its capacity to hold moisture, that air is "saturated" or has 100 percent relative humidity.

When air is cooled beyond its saturation point, it releases the moisture it's holding in the form of clouds, fog, dew, rain, snow, and so on. The temperature at which this happens is called the "condensation point." The condensation point varies depending on the amount of water vapor in the air and the temperature of that air. If a body of air contains a great deal of water, condensation can occur at sixty-eight degrees Fahrenheit, but if the

Warm Front

Cold Front

≫ Frontal lifting ≫

Convective lifting: » Altocumulus. Mid-level layered heap cloud with many convective cells. These clouds are a result of slow lifting that is common ahead of an advancing cold front. The thinner regions between the cells correspond to sinking air with the thick regions to rising air.

air is dry and does not hold much moisture, condensation may not form until the temperature drops to thirty-two degrees or below.

The adiabatic lapse rate is the rate at which air cools as it rises or warms as it descends. This rate varies depending on the moisture content of the air. Saturated (moist) air will warm and cool approximately 3.2 degrees Fahrenheit per 1,000 feet of elevation. Dry air will warm and cool approximately 5.5 degrees Fahrenheit per 1,000 feet of elevation gained or lost.

Cloud Formation

Clouds are excellent indicators of prevailing weather conditions. Their shapes and patterns can be used to forecast weather with little need for additional equipment such as a barometer, wind meter, and thermometer. When air is lifted or cooled beyond its saturation point (100 percent relative humidity), clouds form. There are four ways that air is lifted and cooled beyond its saturation point. They are:

1. Convective Lifting—This occurs when the sun's heat radiating off the earth's surface causes air currents (thermals) to rise straight up and lift air to a point of saturation.
2. Frontal Lifting—A front is formed when two air masses of differing moisture content and temperatures collide. Since air masses will not mix, warmer air is forced aloft over the colder air mass. There it's cooled and

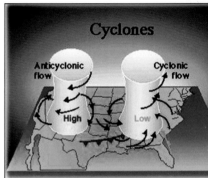

≫ Cyclonic lifting

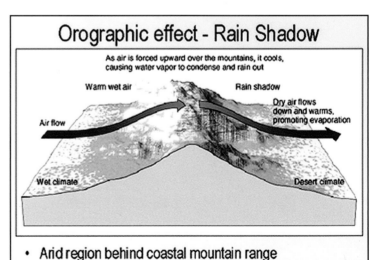

Orographic lifting »

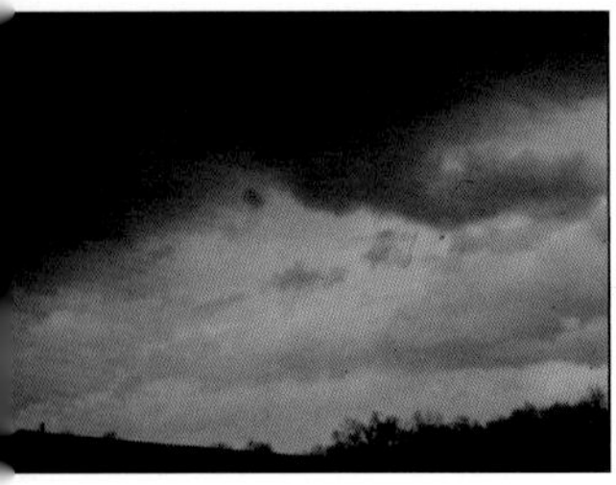

Nimbostratus clouds

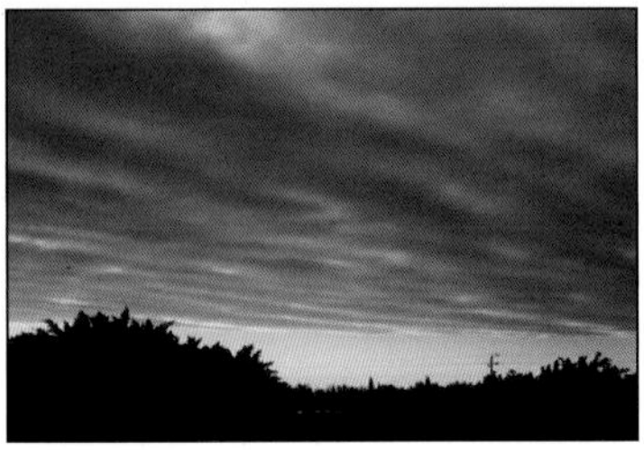

Stratus clouds

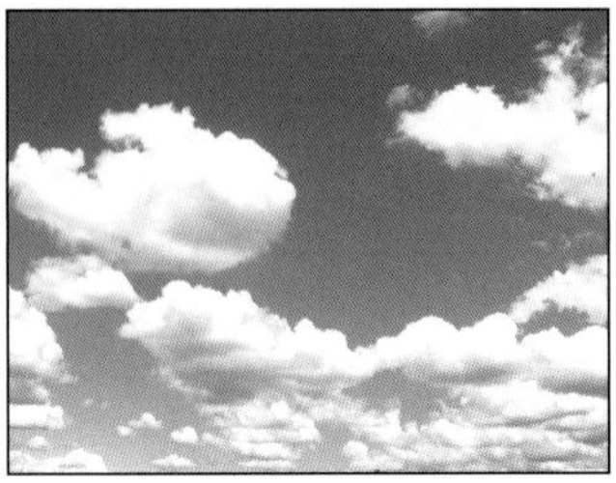

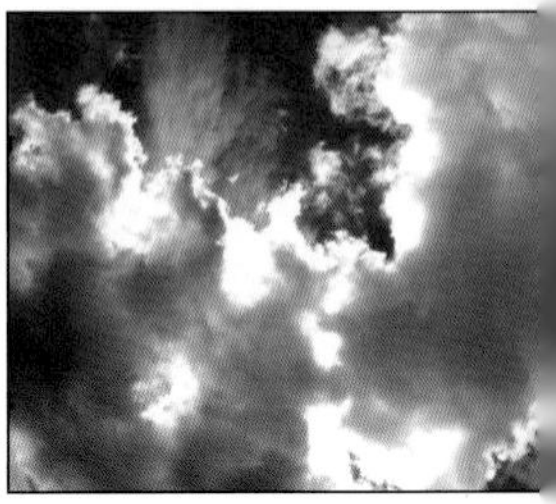

Cumulus clouds

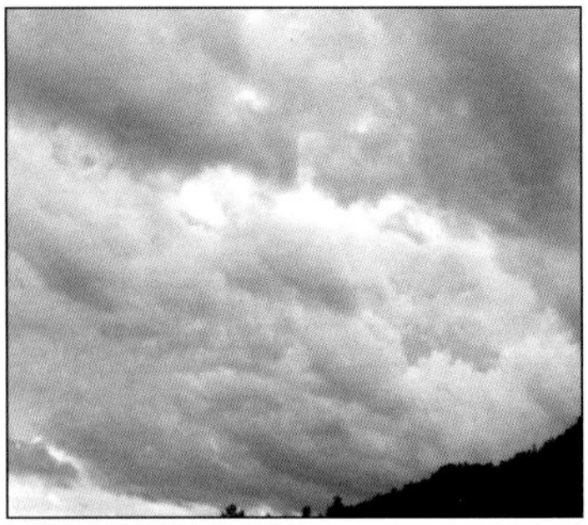

Stratocumulus clouds

can reach its saturation point. Frontal lifting creates the majority of precipitation.

3. Cyclonic Lifting—An area of low pressure pulls air into its center in a counterclockwise direction. Once this outside air reaches the center of the low pressure, it has nowhere to go but up. The air continues to lift until it reaches the saturation point.
4. Orographic Lifting—This occurs when an air mass is pushed up and over a block of higher ground such as a mountain. Air is cooled due to the adiabatic lapse rate until the air's saturation point is reached.

Types of Clouds

Clouds are signposts of the weather. They can be classified by height or appearance, or even by the amount of area covered vertically or horizontally. Clouds fall into five categories: low-, mid-, and high-level clouds; vertically developed clouds; and less common clouds.

Cloud chart

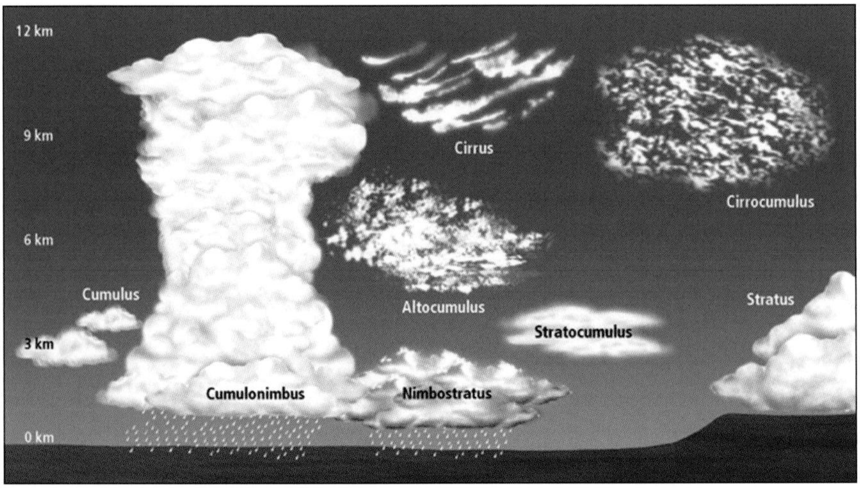

Low-Level Clouds

Low-level clouds (0 to 6,500 feet) are either cumulus or stratus. Low-level clouds are mostly composed of water droplets since their bases lie below 6,500 feet. When temperatures are cold enough, these clouds may also contain ice particles and snow.

Precipitating low-level clouds are either nimbostratus or stratocumulus

Nimbostratus clouds are dark, low-level clouds accompanied by light to moderate precipitation. They block visibility of the sun or moon, which distinguishes them from mid-level altostratus clouds. Because of the fog and falling precipitation commonly found beneath and around nimbostratus clouds, the cloud base is typically extremely diffuse and difficult to accurately define.

Stratocumulus clouds generally appear as a low, lumpy layer of clouds that is sometimes accompanied by weak precipitation. Stratocumulus vary in color from dark gray to light gray and may appear as rounded masses with breaks of clear sky in between.

Low-level clouds can be identified by their height above nearby surrounding relief of known elevation. Most precipitation originates from low-level clouds, because rain or snow from higher clouds usually evaporates before it reaches the ground. Low-level clouds usually indicate impending precipitation, especially if the cloud is more than 3,000 feet thick. Clouds that appear dark at their bases are more than 3,000 feet thick.

⌃ **Altocumulus clouds**

⌃ **Altostratus clouds**

Mid-Level Clouds

Mid-level clouds (between 6,500 to 20,000 feet) are identified by the prefix "alto." Because of their elevated height, middle clouds appear less distinct than low clouds. Alto clouds with sharp edges are warmer because they are composed mainly of water droplets. Cold clouds, composed mainly of ice crystals and usually colder than minus 30 degrees F, have distinct edges that grade gradually into the surrounding sky. Middle clouds usually indicate fair weather, especially if they rise over time. Middle clouds that are descending indicate potential storms, though they are usually hours away. There are two types of mid-level clouds, altocumulus and altostratus clouds.

Altocumulus clouds can appear as parallel bands or rounded masses. Typically a portion of an altocumulus cloud is shaded, a characteristic that distinguishes it from a high-level cirrocumulus. Altocumulus clouds usually form in advance of a cold front. The presence of altocumulus clouds on a warm humid summer morning suggests that thunderstorms are likely to form later in the day. Altocumulus clouds that are scattered rather than even in a blue sky are called "fair weather" cumulus and indicate the arrival of high pressure and clear skies.

Altostratus clouds are often confused with cirrostratus. The one distinguishing feature is that a halo is not observed around the sun or moon. With altostratus, the sun or moon is only vaguely visible and appears as if it were shining through frosted glass.

⋩ **Cirrus clouds**

⋩ **Cirrostratus**

High-Level Clouds

High-level clouds (more than 20,000 feet above ground level) are usually frozen, indicating air temperatures below minus thirty degrees Fahrenheit, with a fibrous structure and blurred outlines. The sky is often covered with a thin veil of cirrus that partly obscures the sun or, at night, produces a ring of light around the moon. The arrival of cirrus indicates moisture aloft and the approach of a traveling storm system—which means that precipitation is often twenty-four to thirty-six hours away. As the storm approaches, the cirrus thickens and lowers, becoming altostratus and eventually stratus. Temperatures are warm, humidity rises, and winds become southerly or southeasterly. The two types of high-level clouds are cirrus and cirrostratus.

Cirrus clouds, typically found at altitudes greater than 20,000 feet, are the most common high-level clouds. They're composed of ice crystals that form when supercooled water droplets freeze. Cirrus clouds generally occur in fair weather and point in the direction of air movement at their elevation. They can form a variety of shapes and sizes from nearly straight to comma-like to all tangled together. Extensive cirrus clouds indicate an approaching warm front.

Cirrostratus clouds are sheet-like, high-level clouds composed of ice crystals. They're relatively transparent, can cover the entire sky, and be up to several thousand feet thick. The sun or moon can be seen through cirrostratus. Sometimes the only indication of cirrostratus clouds is a halo around the sun or moon. Cirrostratus clouds tend to thicken as a warm front approaches, signifying an increased production of ice crystals. As a result, the halo gradually disappears and the sun or moon becomes less visible.

Vertical-Development Clouds

Clouds with vertical development can grow to heights in excess of 39,000 feet and release incredible amounts of energy. The two types of clouds with vertical development are fair-weather cumulus and cumulonimbus.

Fair-weather cumulus clouds have the appearance of floating cotton balls and have a lifespan of five to forty minutes. Known for their flat bases and distinct outlines, fair-weather cumulus exhibit only slight vertical growth, with the cloud tops designating the limit of the rising air. Given suitable conditions, however, these clouds can develop into towering cumulonimbus clouds associated with powerful thunderstorms. Fair-weather cumulus clouds are fueled by bubbles of buoyant air known as "thermals" that rise from the earth's surface. As the air ascends, the water vapor contained with it cools and condenses forming droplets.

Young fair-weather cumulus clouds have sharply defined edges and bases while the edges of older clouds appear more ragged because of erosion. Evaporation along the cloud edges cools the surrounding air, making it heavier and producing a sinking motion outside the cloud. This downward motion inhibits further convection and growth of additional thermals from below, which is why fair-weather cumulus typically have expanses of clear sky between them. Without a continued supply of rising air, the cloud begins to erode and eventually disappears.

Cumulonimbus clouds are much larger and more vertically developed than fair-weather cumulus. They can exist as individual towers or form a line of towers called a squall line. Fueled by vigorous convective updrafts, the tops of cumulonimbus clouds can reach 39,000 feet or higher. Lower levels of cumulonimbus clouds consist mostly of water droplets. At higher elevations, where the temperatures are well below freezing, ice crystals dominate the composition. Under favorable conditions, harmless fair-weather cumulus clouds can quickly develop into large cumulonimbus associated with powerful thunderstorms known as supercells. Supercells are large thunderstorms with deep rotating updrafts and can have a lifetime of several hours. Supercells produce frequent lightning, large hail, damaging winds, and tornadoes. These storms tend to develop during the afternoon and early evening when the effects of the sun's heat are strongest.

Other Cloud Types

Clouds that don't fit into the previous four groups include orographic clouds, lenticulars, and contrails.

Orographic clouds develop when air is forced upward by the earth's topography. When stable air encounters a mountain, it's lifted upward and cools. If that air cools to its saturation temperature during this process, the water vapor within condenses and becomes visible as a cloud. Upon reaching the mountaintop, this heavy air will sink down the other side, warming as it descends. Once the air returns to its original height, it has the same buoyancy as the surrounding air. However, the air does not stop immediately because it still has momentum carrying it downward. With continued descent, the air becomes warmer, causing the surrounding air to accelerate back upwards towards its original height, creating lenticular clouds.

Lenticular clouds are cloud caps that often form above pinnacles and peaks, and usually indicate higher winds aloft. Cloud caps with a shape similar to a flying saucer indicate extremely high winds (more than forty knots). Lenticulars should always be watched with caution. If they grow and descend, expect bad weather.

Contrails are clouds that are formed by water vapor being inserted into the upper atmosphere by the exhaust of jet engines. Contrails evaporate

≈ **Orographic clouds**

≈ **Lenticulars**

≈ **Contrails**

⌃ **Lightning**

rapidly in fair weather. If it takes longer than two hours for contrails to evaporate, that indicates impending bad weather (usually about twenty-four hours prior to a front).

Cloud Interpretation

Cloud cover always appears greater on or near the horizon, especially if the sky is covered with cumulus clouds, since the observer is looking more at the sides of the clouds rather than between them. Cloud cover estimates should be restricted to sky areas more than forty degrees above the horizon—in other words, to the local sky.

Fronts

Fronts occur when two air masses of differing moisture content and temperature meet. One indicator of an approaching front is the progression of the clouds. There are four types of fronts: warm, cold, occluded, and stationary.

A warm front occurs when warm air moves into and over a slower or stationary cold air mass. Because warm air is less dense, it will rise up and over the cooler air. The cloud types seen when a warm front approaches are cirrus, cirrostratus, nimbostratus (producing rain), and fog. Occasionally, cumulonimbus clouds will be seen during the summer months.

A cold front occurs when a cold air mass overtakes a slower or stationary warm air mass. Cold air, because it's denser, forces the warm air up. Clouds observed include cirrus, cumulus, and then cumulonimbus producing a short period of showers.

Cold fronts generally move faster than warm fronts. When they overtake a warm front, warm air is progressively lifted from the surface. Typically the cloud progression will be cirrus, cirrostratus, altostratus, and nimbostratus. Precipitation can be from light to heavy.

A stationary front is a zone with no significant air movement. When a warm or cold front stops moving, it becomes a stationary front. Once this boundary starts moving forward, it becomes either a warm or cold front. When crossing from one side of a stationary front to the other, one will typically experience a noticeable change in temperature and shift in wind direction. Weather is usually clear to partly cloudy along a stationary front.

Temperature

Normally, the temperature drops three to five degrees Fahrenheit for every 1,000 feet gain in altitude when the air is motionless. But when air is moving up a mountain accompanied by condensation (clouds, fog, or precipitation), the temperature of the air will drop 3.2 degrees Fahrenheit for every 1,000 feet of elevation. If air is moving up a mountain and no clouds are forming, the temperature of the air will drop 5.5 degrees Fahrenheit for every 1,000 feet of elevation.

On cold, clear, calm mornings, air temperatures will sometimes rise as you climb. This phenomenon is called temperature inversion. Temperature inversions are caused when mountain air is cooled by ice, snow, and the heat loss of thermal radiation. Cooler, denser air settles into the valleys and low areas. The inversion continues until the sun warms the surface of the earth or a moderate wind causes a mixing of the warm and cold layers. Temperature inversions are common in the mountainous regions of the arctic, subarctic, and mid-latitudes.

At high altitudes, solar heating creates the greatest contrasts in temperature. Since the altitude becomes thinner the higher you go, more direct heat is received than at lower levels, where solar radiation is absorbed and reflected by dust and water vapor. Differences of forty to fifty degrees Fahrenheit can occur between surface temperatures in the shade and surface temperatures in the sun. At sea level this variance is normally seven degrees Fahrenheit. Therefore, at high elevations special care should be taken to avoid sunburn and snow blindness. The clear air at high altitudes also favors rapid cooling at night. Much of the chilled air drains downward due to convection currents so that the differences between day and night temperatures are greater in valleys than on slopes.

Air is cooled on the windward side of the mountain as it gains altitude. This cooling effect (3.2 degrees Fahrenheit per 1,000 feet) occurs more slowly if clouds form due to heat release when water vapor becomes liquid. On the leeward side of the mountain, this heat gained from the condensation on the windward side combines with the normal heating that occurs as the air descends and air pressure increases. Therefore, air and winds on the leeward slope are considerably warmer than on the windward slope. It's important to consider these heating and cooling patterns when planning for mountain travel.

Heat Index Charts

The **heat index** combines air temperature and relative humidity in an attempt to determine the human-perceived equivalent temperature—how hot it feels, termed the "felt" air temperature.

To use the heat index charts, find the appropriate temperature at the top of the chart. Read down until you are opposite the humidity/dew point. The number which appears at the intersection of the temperature and humidity/dew point is the heat index.

Heat Index Chart (Temperature & Dew point)																
Dew point (°F)	Temperature (°F)															
	90	91	92	93	94	95	96	97	98	99	100	101	102	103	104	105
65	94	95	96	97	98	100	101	102	103	104	106	107	108	109	110	112
66	94	95	97	98	99	100	101	103	104	105	106	108	109	110	111	112
67	95	96	97	98	100	101	102	103	105	106	107	108	110	111	112	113
68	95	97	98	99	100	102	103	104	105	107	108	109	110	112	113	114
69	96	97	99	100	101	103	104	105	106	108	109	110	111	113	114	115
70	97	98	99	101	102	103	105	106	107	109	110	111	112	114	115	116
71	98	99	100	102	103	104	106	107	108	109	111	112	113	115	116	117
72	98	100	101	103	104	105	107	108	109	111	112	113	114	116	117	118
73	99	101	102	103	105	106	108	109	110	112	113	114	116	117	118	119
74	100	102	103	104	106	107	109	110	111	113	114	115	117	118	119	121
75	101	103	104	106	107	108	110	111	113	114	115	117	118	119	121	122
76	102	104	105	107	108	110	111	112	114	115	117	118	119	121	122	123
77	103	105	106	108	109	111	112	114	115	117	118	119	121	122	124	125
78	105	106	108	109	111	112	114	115	117	118	119	121	122	124	125	126
79	106	107	109	111	112	114	115	117	118	120	121	122	124	125	127	128
80	107	109	110	112	114	115	117	118	120	121	123	124	126	127	128	130
81	109	110	112	114	115	117	118	120	121	123	124	126	127	129	130	132
82	110	112	114	115	117	118	120	122	123	125	126	128	129	131	132	133

Note: Exposure to full sunshine can increase HI values by up to 15° F

Heat Index Chart (Temperature & Relative Humidity)																
RH (%)	Temperature (°F)															
	90	91	92	93	94	95	96	97	98	99	100	101	102	103	104	105
90	119	123	128	132	137	141	146	152	157	163	168	174	180	186	193	199
85	115	119	123	127	132	136	141	145	150	155	161	166	172	178	184	190
80	112	115	119	123	127	131	135	140	144	149	154	159	164	169	175	180
75	109	112	115	119	122	126	130	134	138	143	147	152	156	161	166	171

70	106	109	112	115	118	122	125	129	133	137	141	145	149	154	158	163
65	103	106	108	111	114	117	121	124	127	131	135	139	143	147	151	155
60	100	103	105	108	111	114	116	120	123	126	129	133	136	140	144	148
55	98	100	103	105	107	110	113	115	118	121	124	127	131	134	137	141
50	96	98	100	102	104	107	109	112	114	117	119	122	125	128	131	135
45	94	96	98	100	102	104	106	108	110	113	115	118	120	123	126	129
40	92	94	96	97	99	101	103	105	107	109	111	113	116	118	121	123
35	91	92	94	95	97	98	100	102	104	106	107	109	112	114	116	118
30	89	90	92	93	95	96	98	99	101	102	104	106	108	110	112	114

Source: *Meteorology for Scientists and Engineers, 2nd edition* by Roland B. Stull

Weather Forecasting

Portable aneroid barometers, thermometers, wind meters, and hygrometers are useful in making local weather forecasts. Reports from any weather service, including USAF, USN, or the National Weather Bureau, are also helpful. But weather reports should be used in conjunction with the locally observable current condition to forecast future weather.

Remember that weather at various elevations may be quite different due to cloud height, temperature, and barometric pressure. Rainy overcast conditions may occur in a lower area, while warmer, clear weather affects the mountains.

The method a forecaster chooses will depend upon the forecaster's experience, the amount of data available, and the level of difficulty that the forecast situation presents.

The five methods of forecasting weather are:

1. Persistence Method. The simplest way of predicting the weather assumes that "today equals tomorrow." If today was hot and dry, the persistence method predicts that tomorrow will be the same.
2. Trends Method. "Nowcasting" involves determining the speed and direction of fronts, high- and low-pressure centers, and clouds and precipitation. For example, if a cold front moves 300 miles during a twenty-four-hour period, we can predict that it will travel 300 miles in another twenty-four hours.

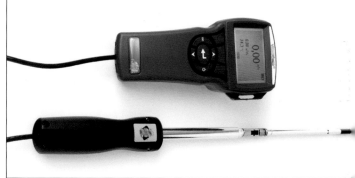

⌃ **Combined air velocity and air temperature meter, thermal anemometer. The telescope probe has a maximum length of 101 cm, suitable for measurements at difficult to reach locations.**

3. Climatology Method. This method averages weather statistics accumulated over many years. This only works effectively when weather patterns are similar from one year to the next.

4. Analog Method. This method examines a particular day's forecast and recalls a day in the past when the weather looked similar. This method is problematic because finding a perfect analogy is difficult.

5. Numerical Weather Prediction. This method uses computers to analyze all weather conditions and is the most accurate of the five methods.

Recording Data

An accurate observation is essential in noting trends in weather patterns. A minor shift in the winds may signal an approaching storm. Establish wind direction as a magnetic direction from which the wind is blowing. Measure wind speed in knots. Use the chart below to determine how fast the wind is blowing.

Speed

(kph)	Effects of Wind
0-1	Calm. No wind; smoke rises straight up.
1-3	Light Air. Smoke moves sideways a little.
4-7	Light Breeze. Wind felt on face; leaves rustle; wind vanes move.
8-12	Gentle Breeze. Leaves and twigs in constant motion; wind moves small flags.
13-18	Moderate Breeze. Wind raises dust and loose paper; small branches move.
19-24	Fresh Breeze. Small trees begin to sway; very small waves on lakes.
25-31	Strong Breeze. Large branches in motion; telephone wires whistling.
32-38	Moderate Gale. Whole trees are in motion; hard to walk against wind.
39-46	Fresh Gale. Wind breaks twigs off trees; cannot walk in wind.
47-54	Strong Gale. Wind damages some houses; awnings blown away.
55-63	Whole Gale. Rarely experienced; trees uprooted; major damage to homes.
64-72	Storm. Lots of damage.
73+	Hurricane. Very rare; lots of damage.

If an anemometer is available, assess speed to the nearest knot. If no anemometer is available, estimate wind speed according to the way objects such as trees, bushes, and tents are affected.

Observe the farthest visible major terrain or man-made feature and determine the distance using any available map. Include any precipitation or obscuring weather. The following are examples of present weather:

- Rain—continuous and steady precipitation that lasts at least one hour.
- Rain showers—short-term and potentially heavy downpours that rarely last more than one hour.
- Snow—continuous and steady frozen precipitation that lasts at least one hour.
- Snow showers—short-term and potentially heavy frozen downpours that rarely last more than one hour.
- Fog, haze—obstructs visibility of ground objects.
- Thunderstorms—potentially dangerous storms. Thunderstorms will produce lightning, heavy downpours, colder temperatures, tornadoes, hail, and strong gusty winds at the surface and aloft. Winds commonly exceed thirty-five knots.

Measure total cloud cover in eighths. Divide the sky into eight different sections measuring from horizon to horizon. Count the sections with cloud cover and record the total cloud cover in eighths. (For example, if half of the sections are covered with clouds, total cloud cover is 4/8.) Estimate where the cloud base intersects elevated terrain. If clouds aren't touching terrain, then estimate to the best of your ability.

Assess temperature with or without a thermometer. With a thermometer, if measurement in temperature is in degrees Celsius, you can convert Fahrenheit to Celsius: C = F minus 32 times 0.55. To convert Celsius to Fahrenheit: F = 1.8 times C plus 32.

Example: 41 degrees F−32 x .55 = 5 degrees C

5 degrees C x 1.8 + 32 = 41 degrees F

Without a thermometer, estimate temperature as above or below freezing (0°C), and as close to the actual temperature as possible.

With a barometer or altimeter, measure pressure trend. A high pressure moving in will cause altimeters to indicate lower elevation. A low pressure moving in will cause altimeters to indicate higher elevation.

Note changes or trends in weather conditions.

Deteriorating trends include:

- Marked shifts in wind direction
- Marked wind speed increases
- Changes in obstructions to visibility
- Increasing cloud coverage
- Increased precipitation
- Lowering cloud ceilings
- Marked cooler temperature changes, which could indicate that a cold front is passing through
- Marked increase in humidity
- Decreasing barometric pressure, which indicates a lower pressure system is moving through the area

Improving trends include:

- Steady wind direction, which indicates no change in weather systems in the area
- Decreasing wind speeds
- Clearing of obstructions to visibility
- Decreasing or ending precipitation
- Decreasing cloud coverage
- Increasing height of cloud ceilings
- Temperature changes slowly warmer
- Humidity decreases
- Increasing barometric pressure

12

NAVIGATION

"Not until we are lost do we understand ourselves."

—Henry Thoreau

≈ Mt. Laguna, Calif—Students from Basic Underwater Demolition/SEAL (BUD/S) class 284 participate in a land navigation training exercise. Land navigation familiarizes students with map and compass navigation in the third and final phase of BUD/S training.

A Survival, Evasion, Resistance and Escape (SERE) instructor from the Center For Security Forces teaches students to identify which direction to travel using a map and compass during the land navigation training at a training site in Warner Springs, Calif. ≽

Whether you're a SEAL on a mission or just trekking through the wild, it's always important to plan a primary and a secondary route in to and out of the area. When you plan, be sure to plan for contingencies. Consider the fitness levels and the condition of the people with you as well as the possible problems you might encounter.

Route Planning

Write down the route you plan to take, broken down into stages, and the time you expect to return. The more detailed the better. Leave this information with a trusted person who will be able to contact help if you do not return when planned.

Keep in mind that your speed of travel will depend of several things:

1. Fitness level (the team can only move as fast as the slowest person)
2. Amount of weight the team is carrying
3. Weather conditions
4. Terrain. A large group will always travel more slowly than one or two people.

Nasmith's Rule

W. Nasmith was a Scottish mountaineer in the late nineteenth century who came up with a formula for estimating the time needed to hike in the mountains that is still used today. According to Nasmith's Rule, you should allow one hour for every five kilometers (three miles) and add thirty minutes for every 1,000 feet (three hundred meters) you gain in height. The rule assumes that you're a fit, experienced climber and doesn't allow for rests, bad weather, or descents. Keep in mind that steep descents can slow you down.

Escape Routes

When planning difficult climbs or treks, always plan your escape routes—in other words, an easy way off a mountain, out of the desert or jungle, or to nearby shelter or SAFE (safe area for evacuation) area. Your escape route should be easy to follow even in bad weather and should not be too steep or difficult in case someone in your party is injured.

Survival Navigation

Stay proficient with your map and compass training. Too many people rely solely on a GPS and don't know how to navigate using a map and compass. Everything mechanical will eventually break, and being under a triple canopy in a jungle or not being able to reach a satellite can cause you serious problems.

Global Positioning Systems (GPS)

There's a big difference between the GPS device you might have mounted in your car and a handheld wilderness GPS. Trip GPS devices are great for urban navigation, supplying alternate routes as well as food, accommodations, and activity options.

Wilderness GPSs units are portable handheld devices that go where you go and are intended for off-road use. Portable handheld GPS devices are designed to help users track routes and find their way out of difficult situations not found on standard maps. Some handhelds even include a two-way radio. Others include removable memory cards for the easy download of multiple destination maps.

All GPS receivers measure three things: your position, your speed, and the current time. (More expensive receivers sometimes include a true magnetic compass and barometric altimeter, but these are completely separate instruments that don't use GPS to make their measurements.) All the other features—things like distance, bearing, map location, and even such vital details as sunrise and sunset or the best fishing times—are just calculations your receiver makes from those three measurements.

There are four basic things you should know regarding how to travel with a GPS:

1. Know how to store your current location in GPS memory. This is known as marking a waypoint.
2. Know how to get back to that stored location from wherever you might be.
3. Know how to program into your GPS receiver the coordinates of locations you want to go to. This is known as entering a waypoint.

≽ **Students in the Survival, Evasion, Resistance and Escape (SERE) course plot their bearing and range to navigate en route to a location at a training site in Warner Springs, Calif.**

≽ **Compass**

≈ Map and GPS

≈ SPOT GPS

4. Know how to navigate from one stored waypoint to the next in succession until you get to your destination. This is known as following a route.

If you know how to do those four things, you'll be able to get where you want to go and return safely using a GPS.

EPIRBs

Before you purchase a GPS, you should do some research and learn the features and technologies of the different models. Some such as the McMurdo Fastfind GPS Personal Location Beacon emit an EPIRB (Emergency Position Indicating Radio Beacon). Once activated by the user (or immersed in water for some marine models), the beacon sends a signal to a satellite above the earth with your specific beacon code. A monitoring site will identify the owner and attempt to contact the emergency numbers provided and also contact the U.S. Coast Guard if near water, and local authorities if on land.

SPOT

Once activated, Spot Satellite GPS Messenger with GPS tracking will acquire your exact coordinates from the GPS network, and send that location along with a distress message to a GEOS International Emergency Response Center every five minutes until cancelled.

GEOS is an entire ecosystem of best-of-breed services that encompass security, safety, and reliable communications for those anywhere in the world. The GEOS programs are delivered through a fusion of state-of-the-art technology with the unparalleled experience of the very best and most respected specialists in the fields of international, personal, and corporate protection; communications; international search and rescue (SAR); and worldwide emergency response.

SPOT's Emergency Response Center next notifies the appropriate emergency responders based on your location and personal information—which may include local police, highway patrol, the Coast Guard, your country's embassy or consulate, or other emergency response or search and rescue teams—as well as notifying your emergency contact persons about the receipt of a distress signal.

SPOT also offers a software package that allows you to set up your personal SPOT profile, which can be managed from a personal computer.

After registering your device, you can configure specific alerts and send out three different types of messages: check-in messages, help requests, and emergency/911 alerts. Each message is sent with information to help find your location, including your latitude and longitude, your device number, the nearest town and how far away it is, and a link to a Google Map with your position located on the map.

TRACMe

A third device available to the public is the TRACMe, which is lightweight and easily activated. When activated, the device transmits a "HELP—EMERGENCY" message every fifteen seconds using Channel 1 of the Family Radio Service radio frequency band (the same frequency used by walkie-talkie radios). The effectiveness of this piece of equipment depends on others knowing that you carry a TRACMe and monitoring your location with an FRS radio. Also, unlike the previous two devices where the signals are received and retransmitted by satellite, the TRACMe issues a line-of-site signal that is unlikely to be detected on the ground much beyond one mile, perhaps two to three miles ground-to-air.

Map and Compass

Modern technology is great. But what happens if you're in the wilderness and you don't have a GPS, EPIRB, SPOT, or a TRACMe? Or if you have an electronic navigational device and the batteries have run out, or it doesn't work? Always have a map and compass and remain proficient at using them together. It is a perishable skill that must be practiced.

⌃ **TracMe**

⌃ **Man in water with GPS**

No GPS, No Compass, No Map!

Don't panic. Your situation calls for an innovative approach. With a little cunning, you'll be able to find your way.

The first thing you'll need is an improvised compass. This can be done with a ferrous metal object such as a needle, pin, nail file, razor blade, metal rod, or something similar, and a suspension system (made with a piece of string or long hair). Magnetize or polarize the metal by slowly stroking it in one direction on a piece of silk or by carefully pulling it through your hair.

Another way to polarize metal is to stroke it repeatedly at one end with a magnet. Always rub in one direction only.

If you have a battery and a piece of electric wire, you can polarize the metal electrically. The battery must be a minimum of two volts. Form a coil with the electric wire and touch its ends to the battery's terminals. If the wire isn't insulated, wrap the metal object in a single, thin strip of paper to prevent contact. Repeatedly insert one end of the metal object in and out of the coil.

A student in the Survival, Evasion, Resistance and Escape (SERE) course uses a small flashlight to read his map and compass during a night navigation lesson at a training site in Warner Springs, Calif. ⌃

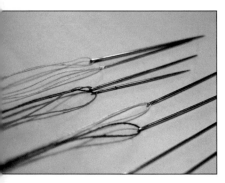

⌃ **Needles and thread**

⌃ **Hiker heading east into the sunrise**

Once your needle is magnetized, suspend it from a piece of nonmetallic string, or float it on a small piece of wood in water. The needle will automatically align itself on a north-south line.

If you have a sewing needle or another thin metal object, a nonmetallic container and the silver tip of a pen, you can construct a more elaborate improvised compass. Cut out the center of the container's lid. Then take your sewing needle and break it in half. One half will form your direction pointer and the other will act as the pivot point. Push the portion used as the pivot point through the bottom center of your container so that it's flush with the bottom and doesn't interfere with the lid. Attach the center of the other portion of the needle on the pen's silver tip using glue, tree sap, or melted plastic. Magnetize one end of the pointer and rest it on the pivot point.

Using the Sun and Shadows

You can also use the earth's relationship to the sun to determine direction. The direction of the rising and setting sun (east and west) is not always accurate due to seasonal variations. But in the Northern Hemisphere, the sun will always appear at due south when at its highest point in the sky or when an object casts no appreciable shadow. And shadows will move clockwise.

In the Southern Hemisphere, the noonday sun will mark due north. And shadows will move counterclockwise in the Southern Hemisphere.

You can also use shadows to determine both direction and time of day with the shadow-tip and watch methods.

Shadow-Tip Method

1. Place a straight, one-meter-long stick on a level spot, into the ground that's free of brush, and onto which the stick will cast a definite shadow.
2. Mark the shadow's tip with a stone or other means. This first shadow mark is always west, everywhere on earth.
3. Wait fifteen minutes until the shadow tip moves a few inches. Mark the shadow tip's new position the same way as you did the first.
4. Draw a straight line through the two marks to create an approximate east-west axis.
5. Now stand with the first mark (west) to your left and the second mark (east) to your right. You are now facing north no matter where you are on earth.

Here's a second shadow-tip method that is more accurate but requires more time:

1. Set up your shadow stick (as you did in the first method) and mark the first shadow in the morning.
2. Use a piece of string to draw an arc through this mark and around the stick.
3. At midday, the shadow will shrink and disappear. In the afternoon, it will lengthen again. Make a second mark at the point where it touches the arc. Draw a line through the two marks to create an accurate east-west line.

The Watch Method

You can also determine direction using a watch. In the Northern Hemisphere, hold the watch horizontal and point the hour hand at the sun. Bisect the angle between the hour hand and the twelve o'clock mark to determine the north-south line. If you have any doubt as to which end of the line is north, remember that the sun rises in the east, sets in the west, and is due south at noon (in the Northern Hemisphere).

This direction will be accurate if you are using true local time, without any changes for daylight savings time. If your watch is set on daylight savings time, use the midway point between the hour hand and one o'clock to determine the north-south line.

If you only have a digital watch, draw a watch on a circle of paper with the correct time on it and use it to determine your direction at that time.

The Moon Method

The moon reflects the sun's light. As it orbits the earth on a twenty-eight-day circuit, the shape of the light it reflects varies according to its position. What we call a new moon, or no moon, is when the moon is positioned on the same side of the earth as the sun. As it moves around the earth, it begins to reflect light from its right side and waxes to become a full moon before waning, or losing shape, to appear as a sliver on the left side.

If the moon rises before the sun has set, the illuminated side will be the west. If the moon rises after midnight, the illuminated side will be the east. This will provide you with a rough east-west reference during the night.

≈ **Sun casting shadows through the trees**

⌃ **Full moon**

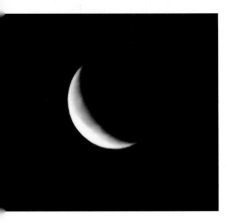

⌃ **Waning moon**

⌃ **Waxing moon**

Here is a rhyme to help you remember if the moon is waxing or waning:

> If you see the Moon at the end of the day,
> A bright Full Moon is on its way.
> If you see the Moon in the early dawn,
> Look real quick, it will soon be gone.

Another way to identify a waxing or waning moon is to determine which side is incomplete. When the moon is illuminated in such a way that the side facing west is incomplete, the moon is waning or getting smaller. If the side to the east is incomplete, the moon is waxing.

Using the Stars

Depending on whether you're located in the Northern or Southern Hemisphere, you will use different constellations to determine your direction north or south.

The Northern Sky

It's important to learn how to identify the constellations Ursa Major (a.k.a., the Big Dipper) and Cassiopeia, which are always visible on a clear night. Use them to locate Polaris, also known as the North Star. The Big Dipper and Cassiopeia are always directly opposite each other and rotate counterclockwise around Polaris, with Polaris in the center. The two stars forming the outer lip of the Big Dipper always point directly to the North Star. Mentally draw a line from the outer bottom star to the outer top star of the Big Dipper's bucket. Extend this line about five times the distance between the pointer stars to find the Polaris (the North Star). Polaris forms part of the constellation Little Dipper.

You can also locate the North Star using the constellation Cassiopeia, which is made up of five stars that form a W-like shape on its side. Extend a line from Cassiopeia's center star to find the North Star. Draw an imaginary line directly to the earth to locate true north.

The Southern Sky

Use a constellation known as the Southern Cross, consisting of five stars, to determine the direction south. The four brightest stars in this constellation form a cross that tilts to one side. The two stars that make up the cross's long axis are the pointer stars. Imagine a distance five times the distance between these stars. The point where this imaginary line ends is in the general direction of south.

Other Methods of Determining Direction

- **Trees**. Growth will be fuller on the side of the tree facing the south in the Northern Hemisphere, and vice versa in the Southern Hemisphere. If you're able to locate several felled trees, look at the stumps. Growth will be more vigorous on the side toward the equator and the tree growth rings will be more widely spaced. On the other hand, the tree growth rings will be closer together on the side toward the poles.

- **Wind**. Wind direction can be helpful where there are prevailing directions and you know what kind they are. (See Chapter 11—Weather).

- **Slopes.** In the Northern Hemisphere, north-facing slopes receive less sun than south-facing slopes and are therefore cooler and damper. In the summer, north-facing slopes retain patches of snow. In the winter, the trees and open areas on south-facing slopes are the first to lose their snow, and the ground snowpack is shallower.

TYVEC Survival Maps —by Wade Chapple

If you can get your hands on a survival map produced by the Joint Personnel Recovery Agency (JPRA), hold on to it. These maps are not only useful for navigation, but each sheet also contains a wealth of survival

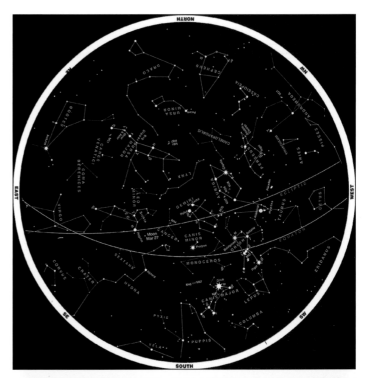

⌃ **Cassiopeia**

information and, as a bonus, they are printed on a special type of rip-resistant and waterproof paper called Tyvec. Unfortunately, the small scale of these maps requires you to carry several sheets if you expect country-wide coverage, and that simply isn't feasible to do on patrol (carry numerous map sheets). So, here's what I recommend you do before heading out:

Select a map sheet that coincides with the principal area in which you will operate, so you are carrying only one map and not the entire (and bulky) series.

Soak the map in a bucket of water and laundry detergent for twenty-four hours or cycle it through a washing machine several times to make the map more pliable and to dampen the noise of the paper in case you're trying to use it in a tactical situation. If you're using the washing machine technique, it takes about four or five cycles to be effective.

Take your survival map to the sewing shop and have grommets installed on all four corners and the centers of all four borders. You can also purchase grommet tools (and grommets) online or at local craft store and do this yourself. Ensure there is at least three-quarters of an inch space between the grommets and the borders of the map.

Optional: Sew in a small square or rectangular-shaped piece of "glint" (reflective) tape for the purpose of emergency signaling.

Lay the map out in the sun to dry or place it in a dryer on low heat.

Once the map is dry, you can fold it to your liking and place it in a pants leg cargo pocket or in one of the outer pouches of your backpack.

If you find yourself in a survival situation, take the map out and read it. You'll find valuable information on edible plants, dangerous animals and insects commonly found in the operational area, instructions for signaling and applying first aid, building fires, preparing food, procuring water, and so on. If it's raining or if you need shade, simply string up the map and sit or lie down underneath it, exactly as you would sit or lie down underneath a poncho. Of course, you can always use the map for its principal purpose, which is navigation!

Methods of Evading Trackers

Evading pursuit involves a combination of speed and misdirection. Your general aim is to put as much distance as possible between you and any pursuit while using misdirection to obscure any tracks you leave.

Tracking a human over rough ground is not easy. But always assume that your pursuers have a massive advantage in man power and resources, with trained dogs, vehicles, and possibly airborne assets to aid in your capture.

Brushing Out Tracks

This is one of the oldest tricks in the book but usually a waste of time. The marks left by brushing away your tracks will tell your pursuers that you're on to them, and use up time and energy. It will also take very little time for a good tracker to rediscover your direction of travel. If the tracker has dogs, it may not slow them at all.

Stream Running

In some cases, running in streams can slow you down; plus, running through even shallow water can sap reserves of strength. In dryer, warmer areas you're more than likely to leave a drip trail. And in tropical areas, it's almost impossible to avoid breaking vegetable as your leave a stream.

If you find a stream that is going in the direction you want, is deep enough for you to float in, and is flowing faster than your walking pace, use it to your advantage. Float feet first and beware of rapids and waterfalls.

Rock to Rock

You can avoid leaving tracks and slow a tracker by jumping from rock to rock. It requires hard stony ground with large rocks that won't shift dangerously when you tread on them—usually found alongside a stream or river—but it can be slow and involves risking a twisted ankle or a fall. Clean the soles of your shoes first to avoid leaving smudges on the rocks as you travel.

Jumping off

This is a technique often taught in the military. Members of the group will jump clear off a trail to circle round and ambush a pursuer.

≈ **Ground pepper**

≈ **Peppermint**

≈ **Garlic**

Backtracking

This involves walking backwards in your own footprints and can be useful when combined with other techniques. A good tracker, however, will notice that as you stride backwards your stride is shorter, your feet farther apart, and your prints deeper and better defined. This works best near a stream or rocks where tracks are harder to follow. Your tracker might not notice when your trail stops, buying you time to get away.

Corner Cutting

This technique works best when you're approaching a road or stream. When you get with 100 meters, alter your course by forty-five degrees (left or right) and continue until you reach the road or stream. Then turn in the same direction again, making sure that your trail leads to the road or stream. Now backtrack to the spot where you met the road or stream and continue along it, trying not to leave any trail. Then leave the road or stream and continue on your way. The idea is to confuse your pursuers and make them think that you saw the road or stream and decided to cut the corner to avoid it.

Evading Dogs

A dog tracks by using its nose, so the way to deal with a dog is to confuse its sense of smell. One way to do that is by dragging a decomposing animal behind you on a length of rope. If you're in an urban area, seek out large groups and strong smells—butcher shops, perfume counters, and so on. Another trick is to get hold of some pepper and pour it behind you.

In rural areas, look for patches of garlic, ransom, and mint. You might try covering a small item of your clothing with stinging nettles. Heavy rain will disperse your scent quickly. If possible, keep downwind of your pursuers.

13

SURVIVAL MEDICINE

"A cheerful heart is good medicine, but a
crushed spirit dries up the bones."

—*Proverbs 17:22*

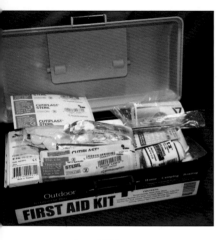

⌃ First aid kit

It goes without saying that medical problems can severely impact your ability to survive in the wilderness. There's nothing more discouraging than being in a survival situation and not having the supplies or, more importantly, the knowledge to be able to treat yourself or a teammate.

That's why it's imperative that basic medical supplies be included in your survival kit, and that you go into the wilderness armed with the understanding of how to maintain optimum health, and recognize and treat certain common medical problems when they appear.

One person with a fair amount of basic medical knowledge can make the difference in many lives. Don't depend on anyone else to have it. If you travel in remote wilderness or dangerous areas, you need to know basic survival medicine.

Survival Requirements for Health Maintenance

To survive, you need water, shelter, food, and some degree of personal hygiene.

Water

Your body loses water constantly during breathing, sweating, urination, and defecation. On average, an adult in an atmospheric temperature of sixty-eight degrees Fahrenheit will lose up to two to three quarts of water a day. Factors such as heat, intense activity, high altitude, and illness can cause the body to lose additional water. All this water must be replaced.

If water isn't replaced, our bodies become dehydrated, which decreases efficiency and increases susceptibility to shock.

The most common signs and symptoms of dehydration are:

- Thirst. (You're at least two percent dehydrated by the time you are thirsty.)
- Dark urine with a strong odor
- Low urine output
- Dark, sunken eyes
- Fatigue
- Emotional instability
- Loss of skin elasticity
- Trench line down center of tongue
- Dry skin and/or lips

Lose five percent of your body's fluid and you will start to feel:

- Thirst
- Nausea
- Weakness
- Anxiety
- Increase in pulse

Lose ten percent of your fluid and you will experience:

- Headaches
- Pins and needles feeling or tingling
- Loss of ability to walk and speak clearly

Lose fifteen percent and you will experience:

- Impaired vision
- Delirium

Lose more than fifteen percent and chances are you will die.

Replace water as you lose it. Even if you don't feel thirsty, drink small amounts of water at regular intervals. When you're under physical or mental stress, you should drink more water.

General guidelines:

- Always drink water when eating. You need water as part of the digestion process.
- Conserve sweat not water. Try to limit sweat-producing activities.
- Ration water sensibly until you find a suitable source.
- To prevent water loss, rest, keep cool, stay in the shade, and seek shelter.
- Do not wait until you run out of water before you look for more.

Salt

Salt is also essential for survival. Not getting enough salt can cause muscle cramps, dizziness, nausea, and fatigue. Having salt tablets or electrolytes in your medical or survival kit can be a big help in a survival situation.

Where to find salt in the wild:

Seawater—Seawater can be diluted with plenty of fresh water to get the salt your body needs. (Never drink pure sea water; doing so can be dangerous to your kidneys.)

Plants—In the United States you can get salt from the root of a hickory tree. You can boil the roots until it evaporates and salt crystals are left.

Foods that naturally have salt—Seafood, carrots, beets, poultry, and most animals.

Salt Licks—A salt lick is a salt deposit that animals lick to get their intake of salt. You can also find them in farming areas set out for cattle.

Food

You can live without food for several weeks, but you will need an adequate supply to prevent your mental and physical abilities from deteriorating. Food replenishes the substances that your body burns and provides energy, vitamins, minerals, salts, and other elements essential to good health.

The two basic sources of food are plants and animals (including fish). In varying degrees both provide the calories, carbohydrates, fats, and proteins needed for normal daily body functions. The average person needs 2,000 calories per day to function at a minimum level.

Plants

A number of plant foods, including nuts and seeds, will provide you with enough protein and oils to function efficiently. Roots, green vegetables, and plants containing natural sugar provide calories and carbohydrates that give the body natural energy.

- Dry plants by wind, air, sun, or fire to retard spoilage so that you can store plant food for when it's needed.
- Plants are easier to obtain than meat.

Animals

Meat provides more protein than plant food. To satisfy your immediate needs, first look for more easily obtained wildlife like as insects, crustaceans, mollusks, fish, and reptiles. These can satisfy your immediate hunger while you prepare traps and snares for larger game.

Personal Hygiene

Cleanliness is an important factor in preventing infection and disease. Poor hygiene can actually reduce your chances of survival.

If possible use a cloth and soapy water to wash yourself daily. Pay special attention to your feet, armpits, crotch, hands, and hair, as these are prime areas for infestation and infection. If you don't have soap, you can either make some yourself, or use ashes or sand. To make your own soap:

1. Cut animal fat into small pieces and cook it in a pot to extract the grease.
2. Add enough water to the pot to keep the fat from sticking, and cook slowly, stirring frequently.

3. After the fat has melted, pour it into a container to harden.

4. Put ashes into a second container with a spout cut near the bottom.

5. Pour water over the ashes and collect the liquid that drips out. This is potash or lye.

6. In your pot mix two parts grease to one part lye.

7. Cook over a fire and boil until the mixture thickens.

8. The result is soap. After it cools, you can use it as a semi-liquid directly from the pot, or pour it into a pan, allow it to harden, and cut it into bars.

If you can't spare water, remove as much of your clothing as is practical and take an air bath by exposing your body to the sun and air for an hour. Watch out for sunburn.

Keep Your Hands Clean

To prevent infection and spread of germs, wash your hands after urinating, defecating, treating the sick and injured, and before handling food, food utensils, or drinking water. Keep your fingernails closely trimmed and clean.

Keep Your Hair Clean

Your hair can become a haven for bacteria, lice, and parasites.

Keep Your Clothing Clean

Clean clothing and bedding will reduce your chances of developing a skin infection. If water is scarce, air clean your clothing by shaking, airing, and sunning it for two hours. If you're using a sleeping bag, turn it inside out after each use and air it out.

Keep Your Teeth Clean

Thoroughly clean your mouth and teeth at least once a day. If you don't have a toothbrush, you can use a chewing stick. Chew one end of the stick to separate the fibers. You can also wrap a clean strip of cloth around your fingers and rub your teeth with it to wipe away food particles. If you don't have toothpaste, brush your teeth with small amounts of sand, baking soda, salt, or soap. Then rinse your mouth with water, saltwater, or willow bark tea.

You can make temporary fillings for cavities by placing candle wax, tobacco, aspirin, hot pepper, toothpaste or powder, or portions of a gingerroot into the cavity. Make sure you clean the cavity thoroughly first.

Take Care of Your Feet

Always break in new shoes before venturing into the wilderness. If you develop a small blister, don't open it. Apply padding around the blister to

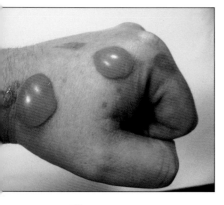

⩘ **Blisters**

relieve pressure and reduce friction. If the blister bursts, treat it as an open wound. Clean and dress it daily and pad around it.

If you develop a large blister, use a sewing-type needle and a piece of clean thread. Clean the outside of the blister, then run the needle and thread through the blister. Detach the needle and leave both ends of the thread hanging out of the blister. The thread will absorb the liquid inside and ensure that the hole doesn't close up so the blister can dry out and heal. Pad around it.

Get Sufficient Rest

It's important to learn how to make yourself comfortable under less than ideal conditions. Take regular rest breaks during your daily activities.

Survival First Aid

In case of an accident or injury, perform a rapid physical exam. Look for the cause of the injury and follow the ABCs of first aid.

A- Airway

Any one of the following can cause airway obstruction, resulting in stopped breathing:

- Foreign matter in mouth or throat that obstructs the opening to the trachea, such as blood, mucous, vomit, broken bones or teeth.
- Face or neck injuries.

ABC: Airway, Breathing, Circulation ⩔

- Inflammation and swelling of mouth and throat caused by inhaling smoke, flames, and irritating vapors or by an allergic reaction.
- "Kink" in the throat (caused by the neck bent forward so that the chin rests upon the chest) may block the passage of air.
- Tongue blocks passage of air to the lungs upon unconsciousness. When an individual is unconscious, the muscles of the lower jaw and tongue relax as the neck drops forward,

Basic Lifesaving Steps

Ref: AFH 36-2218, Vol 1, Vol 2
Use extreme care when treating injuries in a contaminated environment–different rules may apply!

Head tilt, chin lift.

Immediate Steps
When a person is injured:
- Establish an open **Airway** (If possible neck injury, ensure airway opened using the jaw thrust maneuver, do not turn head)
- Ensure **Breathing**
- Stop bleeding to support **Circulation**
- Prevent further **Disability**
  - Immobilize neck injuries
  - Place dressings over open wound
  - Splint obvious limb deformities
- Minimize further **Exposure** to adverse weather

A Airway
B Breathing
C Circulation
D Disability
E Exposure

causing the lower jaw to sag and the tongue to drop back and block the passage of air.

Open an airway and maintain it by using the following steps:

Step 1. Check if the victim has a partial or complete airway obstruction. If he can cough or speak, allow him to clear the obstruction naturally. Stand by, reassure the victim, and be ready to clear his airway and perform mouth-to-mouth resuscitation should he become unconscious. If his airway is completely obstructed, administer abdominal thrusts until the obstruction is cleared.

Step 2. Using a finger, quickly sweep the victim's mouth clear of any foreign objects, broken teeth, dentures, and so on.

Step 3. Using the jaw thrust method, grasp the angles of the victim's lower jaw and lift with both hands, one on each side, moving the jaw forward. For stability, rest your elbows on the surface on which the victim is lying. If his lips are closed, gently open the lower lip with your thumb.

Step 4. With the victim's airway open, pinch his nose closed with your thumb and forefinger and blow two complete breaths into his lungs. Allow the lungs to deflate after the second inflation and:

⌃ **First aid kit, closed**

- *Look* for his chest to rise and fall.
- *Listen* for escaping air during exhalation.
- *Feel* for flow of air on your cheek.

Step 5. If the forced breaths do not stimulate spontaneous breathing, maintain the victim's breathing by performing mouth-to-mouth resuscitation.

Step 6. There is danger of the victim vomiting during mouth-to-mouth resuscitation. Check the victim's mouth periodically for vomit and clear as needed.

B- Breathing

Look, listen, and feel for breathing. Normal breathing rates are between twelve and sixteen breaths per minute.

C- Circulation

Once oxygen can be delivered to the lungs by a clear airway and efficient breathing, there needs to be blood circulation to deliver it to the rest of the body. In case of severe injury, check the patient for bleeding. Severe bleeding from any major blood vessel in the body is extremely dangerous. The loss of less than one quart of blood will produce moderate symptoms of shock. The loss of two quarts will produce a severe state of shock that places the body in extreme danger. The loss of three can be fatal.

⌄ **First aid kit, open**

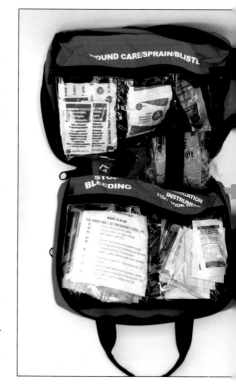

For other cases of severe bleeding:

1. **Have the injured person lie down and cover the person to prevent loss of body heat.** If possible, position the person's head slightly lower than the trunk or elevate the legs and elevate the site of bleeding.

2. **Remove any obvious dirt or debris from the wound.** Don't remove large or more deeply embedded objects. Your principal concern is to stop the bleeding.

3. **Apply pressure directly on the wound until the bleeding stops.** Use a sterile bandage or clean cloth and hold continuous pressure for at least twenty minutes. Maintain pressure by binding the wound tightly with a bandage or clean cloth and adhesive tape. Use your hands if nothing else is available.

4. **Don't remove the gauze or bandage.** If the bleeding continues and seeps through the gauze or other material you are holding on the wound, don't remove it. Instead, add more absorbent material on top of it.

5. **If the bleeding doesn't stop with direct pressure, apply pressure to the artery delivering blood to the area.** Pressure points of the arm are on the inside of the arm just above the elbow and just below the armpit. Pressure points of the leg are just behind the knee and in the groin. Squeeze the main artery in these areas against the bone. Keep your fingers flat. With your other hand, continue to exert pressure on the wound itself.

6. **Immobilize the injured body part once the bleeding has stopped.** Leave the bandages in place and get the injured person to a hospital as soon as possible.

7. **If the wound is abdominal and organs have been displaced, don't try to push them back into place**—cover the wound with a dressing.

Insect Stings

Every country in the world, unless nearly permanently below zero degrees F, has its quota of biting/stinging insects. In some areas, such as tropical rain forests and deserts, the amount of such pests can be staggering.

Insect stings can cause problems in several ways:

1. They can cause high levels of pain, discomfort, and irritation, which can lead to infection.

2. Many insects can be carriers for infectious disease such as malaria, which is on the increase in many countries.

≽ Jungle leg splint

Mosquito ≽

3. Insect bites or stings can cause an allergic reaction through anaphylaxis (anaphylactic shock) or through a toxic overload due to the shear amount of venom, as with many bee or wasp stings. This can result in the airway closing or in respiratory failure.

4. Some insects produce considerable levels of blood loss as they inject a natural anticlotting agent into the bite to keep the blood flowing freely. This can continue after the creature has stopped feeding, since some insects also use a local anesthetic that makes the victim unaware that he is bleeding, as can be the case with the rather unpleasant camel spider. This can be a serious problem if the victim has suffered multiple bites while sleeping.

≽ **Tick**

Treatment

If you're aware that you or a teammate has a serious allergy to insect bites, you should always carry an EpiPen injector (which contains a measured dose of epinephrine made from adrenaline). If an EpiPen injector isn't available, the best you can do is to carefully remove any stinger (or the head and jaws in the case of bulldog ants) left in the wound using tweezers. Be careful to grip the stinger below the poison sack so that extra venom isn't injected into the wound. Then clean the area to prevent infection and apply a cold compress to reduce swelling and any bleeding.

Ticks

Ticks are small blood-sucking insects common in many parts of the world, especially around livestock or grazing animals. The best way to protect yourself is to make sure you are well covered with clothing. After walking through grassy or woodland areas, check your skin and clothing.

If bitten, the risk of infection is high. To remove ticks try to grip the head, which will be very close to the skin, and lever it out with a gentle rocking motion. If parts are left in the wound, remove them and clean the wound carefully and cover to prevent infection.

Leeches

These blood-sucking worm-like creatures can be found in most countries and like moist, humid, and wet conditions commonly found in the jungle. They attach themselves to their victims by sliding off vegetation or floating in water. As they feed, leeches swell in size as they become bloated with blood. In tropical areas, it's best to check for leeches daily, as they can attach to the backs of legs and knees and other hard to reach places.

Leeches cannot be pulled off like ticks because their mouths have a serrated set of teeth. To remove, touch them with a lit or smoldering stick

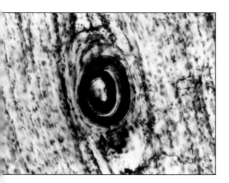

« **Roundworm:** Rats and rodents are primarily responsible for maintaining the endemicity of this infection. Animals, such as pigs or bears, feed on infected rodents or meat from other animals. Humans are accidentally infected when eating improperly processed meat of these carnivorous animals (or eating food contaminated with such meat).

or sprinkle with salt, and they will shrivel and drop off. Once this is done, clean the wound carefully.

Worm Parasites

A wide variety of parasites are common in the poorer areas of the world. Infection is usually caused by swimming in water which has been polluted with sewage. If left untreated, worm parasites can cause great discomfort and reduce strength and stamina.

Roundworm

A pink or white worm, which can grow up to thirty centimeters (twelve inches) in length, and is spread by contact with feces. Early signs can include coughing with some blood as the young worms travel through the blood stream to the lungs, stomach pains, and constipation. People infected with roundworm are usually treated with drugs containing mebendazole or piperazine. If these medications aren't available, mix equal amounts of papaya milk, sugar, and honey, and drink daily.

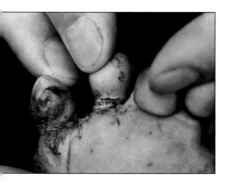

Hookworm: A hookworm infection is also known as "ground itch." Usually the first sign of infection is itching and a rash at the site where skin touched contaminated soil or sand, which occurs when the larvae penetrate the skin, followed by anemia, abdominal pain, diarrhea, loss of appetite, and weight loss. ≽

Threadworm

These one-centimeter-long worms gather around the anus and cause itching. As the victim scratches, these worms travel under the fingernails and into the digestive system via the mouth either directly by hand-to-mouth contact or via food. They're also treated with mebendazole or piperazine, but can be eliminated by drinking garlic crushed into hot water once a day for several weeks.

Hookworm

Hookworm is a particularly nasty worm about one centimeter long that can enter the body in a variety of ways—through infected water or by boring through the skin, especially the soft skin around the feet. Once in the blood stream they travel to the lungs and stomach. Hookworms are very dangerous because they drain strength and stamina and leave their victims vulnerable to pneumonia. To get rid of them completely, you will need medical treatment. But in the short term, iron-rich foods such as red meat and iron tablets will help.

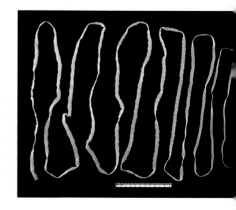

Tapeworm: Humans become infected by ingesting raw or undercooked »
infected meat. The tapeworm can survive for years, attaching to and residing
in the small intestine.

Tapeworm

The tapeworm can reach several meters in length inside the body and normally infects its victim through poorly cooked meat. The effects are generally mild and sometimes go undetected. But pork tapeworms can cause serious damage to the victim's brain and result in death. Infection is usually detected when a small part of worm is found in the victim's clothes or feces. Treatment requires medical attention.

Blood Flukes

Blood flukes are worm-like creatures that enter the body through broken skin or contaminated water. Some can cause bilharzia disease, which results in blood in the urine or diarrhea and is treated with metrifonate or oxamniquine. An estimated 200 million people become infected with blood flukes each year. They're common throughout all of Africa, some islands in the Caribbean, Southeast Asia, and South America.

Amoeba

Amoebas are microscopic organisms that enter the body through infected drinking water. They're so small that few filters are effective in eliminating them. Boiling and other methods of treating drinking water are recommended. They cause amoebic dysentery, which is bad diarrhea with periods of constipation and blood loss. With all cases of diarrhea, it's important to stay hydrated and reduce physical activity.

Rabies

Rabies is a deadly viral disease that is fatal in humans unless the victim receives prompt medical attention. Fortunately, it's fairly rare in the United States. The rabies virus is transmitted through the saliva of infected animals. This happens most often from a bite, although it can also occur from a scratch. Very rarely is it transmitted from saliva contact with broken skin or mucous membranes or from inhalation of aerosolized bat feces. Although the rabies virus can infect just about any mammal, it is most frequently found in raccoons, skunks, bats, foxes and coyotes, and occasionally in cattle and unvaccinated cats and dogs.

If you are bitten or scratched by an animal that you suspect may be rabid, immediately cleanse the wound thoroughly to flush out the virus. Capture the animal if you can do so safely to have it tested for rabies by your local health department. Try not to damage the head since the brain has to be tested for the presence of the rabies virus.

Snake Bites

In most cases, snakes will hear you coming and will move away before you ever know they're around. Also, twenty-five percent of snake bites are dry—which means that the snake hasn't had enough time to load its venom. So even though a bite may draw blood, it does not mean that venom has been injected.

In the event of a poisonous snake bite, an immediate burning pain will spread up the arm or leg that is bitten and swelling will begin within a few minutes to a couple of hours. The affected area may remain swollen, stiff, and bruised for weeks.

Treatment

Elevate the bite if it is on a leg, or use a sling for an arm bite.

Seek medical attention as soon as possible; don't panic.

Antivenin can be administered as much as twenty-four hours after a bite. Remember what the snake looked like to help identify the right antivenin.

14

SURVIVAL KITS

"Luck is where the crossroads of preparation
and opportunity meet."

—*Seneca*

What to pack in your survival kit when your life might depend on it? The answer to this question depends on several contingencies—where you're going, the terrain, climate, the amount you can comfortably carry, and the expected length of the trip.

As you put together your survival kit, consider the different types of products on the market. Know how to use everything in the kit.

U.S. Military Survival Kits

The U.S. military has several types of basic survival kits, which are issued primarily to aviators. There are kits for cold climates, hot climates, and overwater. There is also an individual survival kit that contains a general packet and a medical packet. The cold climate, hot climate, and overwater kits are held in carrying bags and normally stowed in the aircraft's cargo/passenger area. The army's survival kits contain the following:

Cold Climate Kit

- Attaching strap
- Compressed trioxane fuel
- Ejector snap
- First aid kit
- Food packets
- Frying pan
- Illuminating candles
- Insect head net

⪢ **Fishing hook**

- Kit, inner case
- Kit, outer case
- MC-1 magnetic compass
- Packing list
- Plastic spoon
- Pocket knife
- Poncho
- Saw/knife blade
- Saw-knife-shovel handle
- Shovel water bag
- Signaling mirror
- Sleeping bag
- Smoke, illumination signals
- Snare wire
- Survival fishing kit
- Survival Manual (AFM 64-5)
- Survival Utensils
- Waterproof matchbox
- Wood matches

⮛ **Utensils**

Hot Climate Kit

- Attaching strap
- Canned drinking water
- Compressed trioxane fuel
- Ejector snap
- First aid kit
- Fishing tackle kit
- Food packets
- Frying pan
- Kit, inner case
- Kit, outer case
- Kit, packing list
- Insect head net
- MC–1 magnetic compass
- Plastic spoon
- Plastic water bag
- Plastic whistle
- Pocket knife
- Reversible sun hat
- Signaling mirror

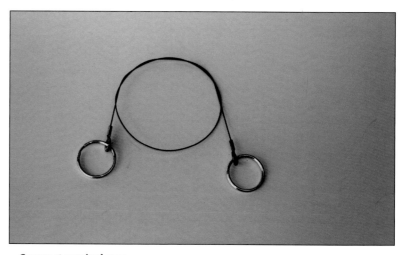

⌃ **Compact survival saw**

- Smoke, illumination signals
- Snare wire
- Sunburn preventive cream
- Survival manual (AFM 64-5)
- Tarpaulin
- Tool kit
- Waterproof matchbox
- Wood matches

Overwater Kit

- Boat bailer
- Compressed trioxane fuel
- First aid kit
- Fishing tackle kit
- Fluorescent sea marker
- Food packets
- Frying pan
- Insect head net
- Kit, packing list
- MC-1 magnetic compass
- Plastic spoon
- Pocket knife
- Raft boat paddle
- Raft repair kit
- Reversible sun hat
- Seawater desalter kit
- Signaling mirror
- Smoke, illumination signals
- Sponge
- Sunburn preventive cream
- Survival manual (AFM 64-5)
- Waterproof matchbox
- Water storage bag
- Wood matches

≈ **Waterproof bag with signal mirror, fire starter, and whistle**

NSN	DESCRIPTION	QTY/UI
1680-00-205-0474	SURVIVAL KIT, INDIVIDUAL SURVIVAL VEST (OV-1), large, SC 1680-97-CL-A07	
1680-00-187-5716	SURVIVAL KIT, INDIVIDUAL SURVIVAL VEST (OV-1), small, SC 1680-97-CL-A07	
	Consisting of the following components:	
7340-00-098-4327	KNIFE, HUNTING: 5 in. lg blade, leather handle, w/sheath	1 ea
5110-00-526-8740	KNIFE, POCKET: one 3-1/16 in. lg cutting blade, & one 1-25/32 in. lg hook blade, w/safety lock & clevis	1 ea
4220-00-850-8655	LIFE PRESERVER, UNDERARM: gas or orally inflated, w/gas cyl, adult size, 10 in. h, orange color, shoulder & chest type harness w/quick release buckle & clip	1 ea
6230-00-938-1778	LIGHT, MARKER, DISTRESS: plastic body, rd, 1 in. w, accom 1 flashtube; one 5.4 v dry battery required	1 ea
6350-00-105-1252	MIRROR, EMERGENCY SIGNALING: glass, circular clear window in center or mirror for sighting, 3 in. lg, 2 in. w, 1/8 in. thk, w/o case, w/lanyard	1 ea
1370-00-490-7362	SIGNAL KIT, PERSONNEL DISTRESS: w/7 rocket cartridges & launcher	1 ea
6546-00-478-6504	SURVIVAL KIT, INDIVIDUAL *consisting of*	1 ea
4240-00-152-1578	GENERAL PACKET, INDIVIDUAL SURVIVAL KIT: w/ mandatory pack bag; 1 pkg ea of coffee & fruit flavored candy; 3 pkg chewing gum; 1 water storage container; 2 flash guards, w/infrared & blue filters; 1 mosquito headnet & pr mittens; 1 instruction card; 1 emergency signaling mirror; 1 fire starter & tinder; 5 safety	1 ea

Individual survival kit with general and medical packets.

NSN	DESCRIPTION	QTY/UI
	pins; 1 small straight-type surgical razor; 1 rescue/signal/medical instruction panel; 1 tweezer, & 1 wrist compass, strap & lanyard	
6545-00-231-9421	MEDICAL PACKET, INDIVIDUAL SURVIVAL KIT: w/carrying bag; 1 tube insect repellent & sun screen ointment; 1 medical instruction card; 1 waterproof receptacle, 1 bar soap & following items:	1 ea
6510-00-926-8881	ADHESIVE TAPE, SURGICAL: white rubber coating, 1/2 in. w, 360 in. lg, porous woven	1 ea
6505-00-118-1948	ASPIRIN TABLETS, USP: 0.324 gm, individually sealed in roll strip container	10 ea
6510-00-913-7909	BANDAGE, ADHESIVE: flesh, plastic coated, 3/4 in. w, 3 in. lg	1 ea
6510-00-913-7906	BANDAGE, GAUZE, ELASTIC: white, sterile, 2 in. w, 180 in. lg	1 ea
6505-00-118-1914	DIPHENOXYLATE HYDROCHLORIDE AND ATROPINE SULFATE TABLETS, USP: 0.025 mg atropine sulfate & 2.500 mg diphenoxylate hydrochloride active ingredients, individually sealed, roll strip container	10 ea
6505-00-183-9419	SULFACETAMIDE SODIUM OPHTHALMIC OINTMENT, USP: 10 percent	3.5 gm
6850-00-985-7166	WATER PURIFICATION TABLET, IODINE: 8 mg	50 ea
	VEST, SURVIVAL: nylon duck	1 ea
8415-00-201-9098		1 ea
	large size	
8415-00-201-9097		1 ea
	small size	
8465-00-254-8803	WHISTLE, BALL: plastic, olive drab w/lanyard	1 ea

Individual survival kit with general and medical packets (continued).

The Basic Navy SEAL "Layout"

U.S. Navy SEALs separate their gear into three categories, or lines. The gear list changes according to area of operation, terrain, climate, enemy situation, support assets, and so on.

First line gear contains the everyday essentials needed for immediate short-term survival. SEALs do not go anywhere without their first line gear. Cammies, weapons, maps, compass, knife, 550 cord, and watches are all typically included in first line gear. If the second and third lines of equipment are abandoned, the first line gear will give the operator enough gear for short-term survival.

Second line gear contains necessary extras carried in load-bearing equipment (LBE) or tactical vests. It's gear that is quickly available should the need arise. Equipment such as ammunition, grenades, water and purification tablets, and medical supplies are all considered part of second line gear. Second line gear is always carried or worn when working (operating).

Third line gear is made up of supplies needed for a mission that aren't as critical for immediate use. Radios and batteries, claymore mines, ponchos, water filters, and night vision goggles are all included in third line gear. Corpsmen fit their medical gear into this category. A SEAL is never too far away from his third line gear. In urban areas, it may be kept in a rucksack in the trunk of a vehicle; while berthing, it may be at the foot of the cot or bed.

Ideally a SEAL is never in a situation where he does not have all of his gear. However, if he is caught in a situation when he must travel light, he may leave behind his third line gear.

 Pliers

Examples of first line gear:

- Wrist compass
- Small LED (red) flashlight
- Surefire (or other compact & bright light)
- Swiss Army knife
- Leg holster
- Sidearm
- Two spare magazines
- Small water container
- Large knife on leg holster (opposite side)
- Compact survival kit
- Fishing hooks, line, sinkers

≋ **Pocket knife**

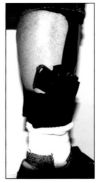

≋ **Leg holster**

≋ **Survival knife with compass top**

- Fire starter and waterproof matches (inside waterproof container)
- Water purification tablets
- Parachute cord
- Signal mirror
- Whistle

Examples of second line gear

- Assault rifle
- Four spare magazines
- Flares
- Carabiners
- Spare flashlight batteries (enough for twenty-four hours)
- Water container
- Binoculars
- Body warmer (air-activated pad)
- Poncho with liner
- Several pouches of freeze-dried food
- QuikClot (powder that quickly clots blood)
- Large gauze compress
- Duct tape
- Snakebite kit
- Maps
- Compass
- Fire starter (cotton smeared in Vaseline inside of a pill bottle)
- Lighter

≋ **Flares**

≋ **Dehydrated meal**

≋ **Warmers**

Examples of third line gear

- Fifty feet of parachute cord/550 cord
- Large water container
- Water purifier
- Sleeping bag or bivy sack
- Extra socks
- Ten full freeze-dried meals
- Rain gear
- Batteries (twenty-four-pack per device)
- Flashlight
- Extra ammo (usually 300 rounds per weapon)
- Extra pair of gloves

Customize your gear to fit your personal/operational needs.

Your survival kit should be:

- Lightweight and compact
- With you when you need it
- Packed with equipment you can rely on
- Adjusted for the season and expected weather

Be sure to:

- Periodically check the serviceability of all survival kit components.
- Check that each person in your group carries his own survival kit.
- Build your own survival kit, since commercial kits compromise the quality of the components in order to keep the overall price of the kit down—it's better to build your own.

Basic Survival Kit Considerations

Water

- Water-purifying straw. Good for twenty to twenty-five gallons of water (depending on how contaminated it is). It's not just a filter, but actually treats water with antibacterials.
- Water purification tablets. Each tablet can purify one or two quarts of water, depending on how dirty the water is. You can also crush one water purification tablet and add about a teaspoon of water to make a strong iodine topical solution for treating injuries.

- One-hundred ounce reservoir filled with water
- Five feet of plastic tubing, i.e., fish tank tubing
- Bottle ORS
- aluminum canteen cup

Shelter/Bivy Material

- Tent
- Bivy sack
- Sleeping bag
- Large, heavy-duty orange plastic or Mylar survival bags
- Kydex/painter's tarp (nine feet by seven feet). This thin plastic sheet can be used for a variety of purposes—making a shelter, waterproofing a roof, collecting rainwater, or used as a solar still. It tears easily, so you might want to pack two.
- Survival blanket. Made of durable, tear-resistant polyethylene, these blankets reflect back body heat to keep you warm. Make a frame out of sticks, and stretch the blanket over it; you can also use it as a very large signaling mirror. Also works for waterproofing your shelter, collecting rainwater, or wrapping around yourself as a poncho or shawl.
- Plastic or fabric tube tents
- Sheet plastic
- 550 cord

Fire-Starting Material

- Windproof lighter
- Survival matches
- Metal match with a scraper.
- Magnifying glass
- Magnesium/flint bar fire starter. Shave the magnesium with a knife; collect all the shavings into a pile about the size of a dime. The magnesium ignites with a flame like a blow torch and will burn for several seconds.
- Tinders
- Thirty-six hour candle in a can
- Survival saw

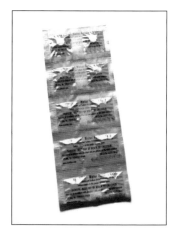

≈ **Water purifier tablets**

≈ **Survival bivy sack**

≈ **Survival blanket**

≈ **Fire starter, whistle, matches, signal mirror**

≈ **Waterproof matches, tape, fire starter sticks**

Signaling Equipment

- Whistle with lanyard
- Signal mirror
- Fluorescent plastic surveyor's tape
- Brightly colored fabric
- Fire starting material
- Light sticks or chemical lights
- Rugged LED strobe/flashlight

Personal Protection

- Whistle with lanyard
- Emergency foil bag/space blanket
- Body warmers
- Additional clothing, including rain gear—for warmth and protection from wind and wet
- Polypropylene balaclava
- Insect head net
- Bug spray
- Sunscreen and sunglasses

First Aid Kit

- Sunscreen and sunglasses
- Wound prep pads
- Soap towelettes
- One-by three-inch adhesive bandages
- Butterfly bandages
- Two-by four-inch flex bandages

- Knuckle bandage
- Eye patch
- two-by-three-inch nonstick pads
- Hydrocortisone cream
- Triple antibiotic ointment
- Burn ointment
- Motrin
- Tylenol
- NoDoz
- Imodium AD
- Safety pins
- Lip balm SPF 15
- Roll one-half-inch medical tape
- Tweezers
- Waterproof pouch, five by seven inches

Medical Trauma Gear

- Trauma dressings
- High-absorbency gauzes
- Tourniquet
- Snakebite kit

Headlamp with spare bulbs and batteries

LED flashlight with spare batteries and bulbs. I recommend a mini LED flashlight with either a white, yellow, green, or blue light. They're great when you need to locate items in the dark, and can also be used for signaling at night. Pack extra batteries, as they take up very little space and last a long time. I recommend putting some electrical tape between each to keep them from discharging. As each battery is about the size of a dime, this takes very little space in the kit. If your light requires a tool to change the batteries, make sure you have one with you.

Food Supplies

- Hard candies
- Nuts and seeds
- Bouillon cubes
- High calorie emergency rations (such as energy bars high in carbohydrates)
- Commercial dried packaged food

- MREs
- Eating utensils/cup/spork
- Thirty feet of fifteen-pound test fishing line, fifty feet of eighty-pound test fishing line, and six fishhooks for fifteen-pound test line Use the lighter line for fishing in rivers and streams. The heavier line is as tough as nails and has thousands of uses—including shelter building, snare making, and for unattended fishing purposes.
- Hooks
- Fishing flies
- Snares

Navigation

- Maps in waterproof containers
- Compass
- GPS
- Five feet orange flag tape (for marking your position)

Miscellaneous Gear

- Knife and/or multitool
- Sewing kit
- Croc-Lock™ clips
- Five feet duct tape wrapped around water containers. Makes it possible for you to repair just about anything—patch holes in tarps, bandage cuts, fix a point to an arrow or spear, and so on.
- 550 paracord (twenty feet). The uses of paracord in a survival situation are almost too numerous to list. Use it for shelter, whether for tying frame members together or for stringing up a tarp between trees. It also works for snares and building other weapons. True paracord is constructed of an outer sheath, which contains seven inner strands. Each inner strand is fifty-pound test, while the outer sheathing is rated at about 200 pounds. If you need more string in an emergency situation, you can remove the inner strands. The outer sheathing can serve well as shoelaces.
- Safety pins. For quick repairs of clothing and other gear. Can also be fashioned into emergency fishhooks.
- Zip ties. Use them for setting up your emergency shelter, among other purposes.
- Ziploc freezer bags. Use as canteens, waterproof storage, and so on.

- Scalpel blades. Come in sterile wrapping, are extremely small, and have a multitude of uses, from medical to skinning and gutting game to whittling.
- MilSpec snare wire. Has a multitude of uses, from snaring to trip wire to hanging food items over a fire for cooking.
- Fifty dollars cash in local currency (or even an extra credit card).

Familiarity with Survival Equipment—by Wade Chapple

Being on your own in a life or death struggle for survival is not the time to familiarize with your equipment, nor is it the time to pull out a handheld GPS only to find it without batteries or with the batteries depleted. It goes without saying that special operators are always intimately familiar with and can effectively maintain the weapons systems they carry.

Unfortunately, I have found that these same special operators do not always have a good understanding of how to operate or maintain the survival equipment packed into their "go" bags. For that matter, unfamiliarity with survival aids and equipment is a common trend I find with all of our clients, both U.S. and foreign.

For example, we were once called to assist in the search for two missing crew members who had ejected from their fighter jet over the mountains of Colombia. We collectively managed to safely recover the two crew members, but during our post-incident analysis, we discovered that the pilot was actually carrying a survival radio with a 406MHz emergency distress beacon, and he did not activate it at all. If he had activated the survival radio, we would have located his position very quickly and reduced his exposure to the elements and to the enemy who was also searching for him.

After questioning the pilot, I learned that he had never received any training on how to operate or maintain the survival radio and that the radio's battery was dead because he had never checked or tested it before flying. The crew members were very fortunate in this particular example, but their plight would definitely have been shortened if the pilot knew how to operate and maintain his survival radio before becoming isolated.

Whatever it is you decide to pack for use during a survival situation, the following guidelines will ensure that those items are useful to you in a time of need:

- Learn how to operate or use all of the survival gear that you pack. If it requires batteries, make sure the batteries are fresh before going to the field and pack extra (sealed) batteries in order to extend the use of battery-operated items during survival situations. While training for survival situations, use the equipment you are carrying to ensure proficiency with the same.

- Before going out to the field, inspect all of your survival gear. Turn on and conduct a function test of anything that is battery operated.

- Try to acquire equipment with common batteries (if a flashlight uses AA batteries, for example, try to acquire a GPS and strobe light that also use AA batteries).

- Swap out items that are damaged.

- Organize your survival gear so you don't have to fumble around at night looking for the item that you need.

- I have found that battery-operated chemlights are better for survival situations than the disposable, one-time-use chemlights. However, I also pack a couple of infrared chemlights that are one-time use but very good for tactical signaling.

15

THE MYSTERY
OF SURVIVAL

"We had suffered, starved and triumphed,

groveled down yet grasped at glory . . .

We had reached the naked soul of man."

—*Sir Ernest Shackleton*

≫ **Candiru fish**

≫ **Maggots**

Juliane Koepcke

On Christmas Eve 1971, a quiet seventeen-year-old high school senior named Juliane Koepcke and her mother boarded LANSA airline Flight 508 from Lima, Peru, to the Amazon jungle city of Pucallpa. She was on her way to meet her father, who ran a research station in the jungle. Twenty-five minutes into the flight, the Lockheed Electra turboprop ran into bad weather. "We ran into heavy clouds and the plane started shaking," Juliane recalled. "Then, to our right, we saw a bright flash and the plane went into a nose dive."

Accident investigators later determined that one the of the plane's fuel tanks was struck by a bolt of lightning. "Christmas presents were flying around the cabin and I could hear people screaming," Juliane said. The plane then broke into pieces, and Juliane passed out. When she gained consciousness, she was still strapped to her seat and alive despite the fact that she had fallen two miles into the jungle canopy.

"Maybe it was the fact that I was still attached to a whole row of seats," she explained later. "I was rotating much like a helicopter and that might have slowed the fall. Also, the place I landed had very thick foliage and that might have lessened the impact." The other ninety-one people aboard Flight 508 died.

Juliane's injuries were relatively minor—she had a broken collarbone, her right eye was swollen shut, she had a concussion, plus she had large gashes on her arms and legs. But she had lost her eyeglasses, and she was alone deep in the Amazon jungle with the thick canopy above her preventing her from signaling for help. Juliane had no food, no tools, no compass, no map, and no means to make a fire. Just herself, surrounded by thousands of species, including some of the most venomous creatures on earth.

But Juliane didn't give up. Instead, she searched through the plane's wreckage, grabbed a few pieces of candy and cake, and started walking through the jungle. She had visited her father's jungle research station before and knew that if she followed water downstream it would likely lead to some form of civilization. The day after the crash, she found a creek and started to follow it downstream. When the creek ran into a larger body of water, she followed that. When the vegetation on the river bank grew too thick, she waded knee-deep through the piranha and candiru-infested water (the murderous candiru fish swims up its victim's urethra and feasts on the internal organs). The going was tough.

"I had a cut in my arm, and after a few days I could feel that there was something in it," Juliane remembered. "I took a look in the hole and saw that a fly had laid her eggs in the hole. It was full of maggots. I was afraid I would lose my arm."

Farther downstream she discovered more wreckage of the plane. "I found another row of victims, with three dead women strapped in. They had landed head-first and the impact must have been so hard that they were buried two feet into the ground."

But she still didn't give up. Instead, she continued to fight her way through snarls of vegetation, swarms of insects, and leeches. She drank the river water, foraged for whatever scraps of food she could find, and waded through streams infested with crocodiles, piranhas, and devil rays. "Sometimes I would see a crocodile on the bank and it would start into the water towards me, but I was not afraid," she said. "I knew that crocodiles don't tend to attack humans."

Starved and exhausted, after ten days of trekking through the jungle Juliane came upon a small boat and a hut. She stayed there, and the next day was discovered by a group of Peruvian lumberjacks. They loaded her into their canoe and paddled seven hours to the nearest town, where a local pilot flew her to a hospital for treatment. Doctors discovered more than fifty maggots in her arm.

She went on to earn a PhD in zoology and now works at the Zoological Center in Munich, Germany. Her survival story remains one of the most amazing feats of human endurance I've ever heard.

Juliane Koepcke, who claims that she isn't a spiritual person, has said that she looks at "logical" reasons for why she survived. Others who have found themselves in similarly harrowing survival situations, however, have attributed their ability to come out alive to other factors.

Frank Smyth

Take British explorer Frank Smyth. In 1933, he almost reached the summit of Mount Everest, which would have made him the first person to do so. The situation was horrendous: freezing temperatures, powerful winds, and blinding, swirling snow. While the rest of his hiking party fell back, Smyth continued to press ahead, coming within 1,000 feet of his goal.

Later, writing in his diary, Smyth remembered how at one point during the ascent, he reached into his pocket, pulled out a slab of Kendel Mint Cake, broke it in half, and turned to give the other half to a companion. But there was no one there. "All the time that I was climbing alone, I had a strong feeling that I was accompanied by a second person," he wrote. "The feeling was so strong that it completely eliminated all loneliness I might otherwise have felt."

The Third Man Syndrome

This is an example of what is known as "The Third Man Syndrome." Also described as the Third Man Factor, it refers to documented cases where an unseen presence or spirit has provided comfort and support in a traumatic situation. In recent years, well-known adventurers such as climber Reinhold Messner and polar explorers Peter Hillary and Ann Bancroft have reported having experiencing this phenomenon. In 2009, John G. Geiger wrote a book called *The Third Man Factor: Surviving the Impossible* (Weinstein Books), which documents many examples.

Sir Ernest Shackleton

One of my greatest sources of inspiration is Sir Ernest Shackleton, the legendary Anglo-Irish explorer. I can't tell you how many times I've thought back to his incredible story, his fortitude, and his never-give-in attitude. He once said, "Never for me the lowered banner, never the last endeavor."

Shackleton's Antarctic expedition of 1914—during which he and his crew endeavored to cross the continent from sea to sea—ultimately failed, but the story of how he and the men with him fought against incredible odds won them many accolades and honors.

After their ship *Endurance* became trapped in ice, Shackleton and his men were forced to make a grueling journey across ice floes, then in lifeboats to Elephant Island, where for six month the main group subsisted on seal meat and blubber.

Shackleton took five men around the island to the north and then across 800 miles of treacherous ocean to South Georgia Island. He then hiked and climbed with two others for thirty-six hours across the island's uncharted interior to a whaling station, with another three months to go before he could safely reach the crew back on Elephant Island. Shackleton later wrote: "I know that during that long and racking march of thirty-six hours over the unnamed mountains and glaciers, it seemed to me often that we were four, not three."

Years later, the poet T.S. Eliot read Shackleton's account of a mysterious "fourth" man and included it in his famous poem, *The Waste Land*.

⚌ **Sir Ernest Shackleton and wife**

⚌ **Sir Ernest Shackleton**

Using poetic license he turned Shackleton's fourth into a third, which is where the phenomenon gets its name.

> Who is the third who walks always beside you?
> When I count, there are only you and I together
> But when I look ahead up the white road
> There is always another one walking beside you
> Gliding wrapt in a brown mantle, hooded
> I do not know whether a man or a woman
> But who is that on the other side of you?

Over the years, the Third Man experience has occurred over and over to people on the edge of death—including 9/11 survivors, mountaineers, divers, polar explorers, prisoners of war, solo sailors, aviators, and astronauts. All have escaped traumatic events only to tell strikingly similar stories of having experienced the close presence of a helper or guardian.

Ron DiFrancesco

At 8:46 AM on September 11, 2001, thirty-seven-year-old Canadian money-market broker Ron DiFrancesco was sitting at his desk on the eighty-fourth floor of the South Tower of the World Trade Center in New York, when a plane struck the North Tower. The world around him changed immediately. As the lights in his office flickered on and off, he saw gray smoke pouring from the North Tower. He watched as people in offices trapped above the ninety-first floor on the North Tower waved desperately for help. Most his fellow workers at Euro Broker started to evacuate the building. Then an announcement came over the South Tower PA system, which stated: "Building Two (the South Tower) is secure. There is no need to evacuate Building Two. If you are in the midst of evacuation, you may return to your office by using the reentry doors on the reentry floors and the elevators to return to your office. Repeat, Building Two is secure . . . ."

DiFrancesco telephoned his wife, Mary, to tell her that he was fine and intended to stay at work. "It was Tower One that was hit," he assured her. "I'm in Tower Two."

Then a friend from Toronto called and told him, "Get the hell out." DiFrancesco decided to heed his friend's warning and started walking toward a bank of elevators when United Airlines Flight 175, traveling at 595 miles an hour, slammed into the South Tower, hitting the building between the seventy-seventh and eighty-fifth floors.

DiFrancesco was thrown against a wall and showered with ceiling panels and other debris. With the building swaying and the fire from over 11,000 pounds of jet fuel spreading, DiFrancesco entered Stairway A. He and others started to descend through the smoke-filled stairwell, lit only by a flashlight carried by a colleague and executive vice-president at Euro Brokers, Brian Clark.

Three flights down, they ran into a heavy woman and a man who were on their way up. "You've got to go up. You can't go down," the woman insisted. "There's too much smoke and flames below."

While Clark and DiFrancesco debated whether to climb to the roof or descend, they heard someone screaming for help. The two men entered the eighty-first floor to try to locate the distressed man, but DiFrancesco was quickly overcome with smoke. He reentered the stairwell and started to climb. But the further he climbed, the more clogged with people the stairwell became. And the emergency fire doors on the landings were locked.

By the time he reached the ninety-first floor landing, DiFrancesco started to panic. He had a wife and two young children who needed him. He had to get out of the building alive. With only one other option, DiFrancesco decided to turn around and go back down. But as he approached the impact zone, the smoke grew thicker and more astringent. At the landing of the eightieth floor, he found that a collapsed wall and flames blocked his descent. He got down on the floor with about a dozen other desperate people and started gasping for air. Some were crying. Others were slipping into unconsciousness. It seemed like he was about to die with them in the stairway, when he heard a voice address him by his first name and say, "Get up, Ron! You can do this!"

He later said that he felt he was being guided by a physical presence that led him down the stairs. After fighting his way through drywall and other debris, he wanted to turn back, but the presence urged him to continue. So Ron DiFrancesco covered his head with his forearms and ran down through three stories of flames. Singed by the flames and in pain, he kept descending. Upon reaching the ground floor, he started to exit when a security guard stopped him and pointed to another, safer exit near Church Street.

Fifty-six minutes after the second plane hit, the floors of the South Tower began to pancake down. DiFrancecso remembered hearing an "ungodly roar" as he approached the exit. He turned and saw an enormous fireball headed toward him, but can't recall what happened next. Days later he awoke in St. Vincent's Hospital in Greenwich Village with lacerations on his head, burns all over his body, and a broken back.

≈ **World Trade Center**

≈ **9/11 survivors**

≈ **9/11 destruction**

By all accounts, he was the last person to escape the South Tower of the World Trade Center before it came down at 9:59 AM, and one of only four people to escape the building from above the eighty-first floor. Ron DiFrancesco doesn't understand why he lived and others didn't. But he has no doubt that it was "an angel" who guided and urged him through the impact zone to safety.

Angel or Hallucination?

Fabled climber Reinhold Messner, who has experienced more than one encounter with the Third Man, believes it's a natural phenomenon produced by the brain. "I think all human beings would have the same or similar feelings if they would expose themselves to such precarious situations," he explained. "The body is inventing ways to let the person survive."

Whether you believe that the Third Man is an angel or some sort of hallucination produced by the mind under extreme stress, there's no doubt that Ron DiFrancesco, Frank Smyth, Sir Ernest Shackelton, Reinhold Messner, and Juliane Koepcke had one thing in common—an indomitable will to survive.

ACKNOWLEDGMENTS

As a Training Officer in the Navy SEAL Teams I had the responsibility and great fortune of instructing SEALs in Desert Survival, Jungle Survival, Arctic Survival, Mountain Survival, Sea Survival and Urban Survival. In the SEAL teams, we not only trained ourselves, but quite often would reach out to other professionals, such as the individuals listed below to instruct us.

In *The Navy SEAL Survival Handbook* I not only included much of the survival information that I instructed in the SEAL teams, but I also reached out to the professionals who live and breathe "survival" on a daily basis. I am truly indebted to these world renowned survival experts who contributed a wealth of information to this book.

Dr. Bruce Jessen

Dr. Jessen, a former leading Psychologist for the Joint Services Survival, Evasion, Resistance, and Escape (SERE) Course at Fairchild *AFB,* WA. He is a world renown, leading authority on the topic of survival during captivity. Dr. Jessen contributed his paper on *RESILIENCE: CAN THE WILL TO SURVIVE BE LEARNED?* Dr. Jessen spent most of his professional career better preparing our military with skills and techniques on how to survive in governmental or non-governmental captivity.

Tony Nester

http://www.apathways.com/media/course-picts/index.html

Tony owns and instructs at Ancient Pathways in Desert Wilderness Survival, Primitive Skills, and Bushcraft, and is a regular contributor to Outside Magazine. Tony has shared his wealth of knowledge with civilians, the Navy SEAL Teams, other members of the Special Operations community, and with many other military organizations. He is the most sought after Desert Survival instructor in North America. Tony provided us with a great deal of information on desert survival issues such as water procurement, flash floods, edible plants, trapping, recommended clothing and foot wear, food procurement and survival kit lists.

David Ayres

www.tate-inc.com

President - TATE, Incorporated

TATE is a leading provider of SERE and personnel recovery training for the US government.

Wade R. Chapple
www.tate-inc.com
As the Managing Director for Latin America at TATE Incorporated, Wade was very helpful in providing information in Survival Evasion, Resistance and Escape training in jungle environments, including creating survival maps, crossing jungle rivers, and protecting clothing from the elements.

Dave Williams, Director of Paddle Asia
http://paddleasia.com
Dave teaches Jungle Survival in Thailand and is a world renowned expert in jungle survival, including edible plants, trapping, survival kits, water procurement, starting fires, cooking, weather, building shelters and navigation.

Additionally, this book would not have been possible without the hard work and determination of Jay Cassell and his professional team at Skyhorse Publishing.

I was very fortunate to collaborate with my very talented co-author and friend Ralph Pezzullo [www.RalphPezzullo.com], as well as Erika Hokanson [www.RefreshMediaResources.com] who collected, sorted and acquired permission to publish photos that added such great value to this project.

PHOTO CREDITS

68: (T) Avanzero; (B) Kevin Connors; • **69:** U.S. Navy photo by Mass Communication Specialist 2nd Class Matt Daniels; • **70:** (All) Don Mann; • **71:** (TL) National Weather Service; (TR) U.S. Federal Gov.; (BR) phestus; • **72:** (T) Figure 31, U.S. Geological Survey; (B) Maria Ly/Skimble.com; • **73:** (T) Serious fun; (B) U.S. Geological Survey;

74: Don Mann; • **75:** Dr. Becky Abell; • **76:** Steve Evans; • **77:** (T) Robespierre; (C) Kevin Connors; (B) Héctor Maffuche, Revista Argentina "Gente y la actualidad" 1974

78: (All) Don Mann; • **79:** Dr. Becky Abell; • **80:** (All) Don Mann; • **82:** (T) Kevin Connors; (B) Scott M. Liddell; • **83:** Phil Kates; • **84:** ricorocks; • **85:** Thomasie; • **87:** (All) Tony Nester;

90: Tony Nester; • **91:** xandert; • **97:** U.S. Navy photo by All Hands Photographer's Mate 1st Class Michael Larson; • **98:** (T) Kcd88; (B) xandert; • **99:** (T) D3designs; (C) Ben Earwicker; (B) dustie; • **100:** (T) Paul David Lewin; (B) cohdra; • **101:** (L) xandert; (R) dieraecherin; • **102:** (T) Kittenpuff1; (B) jeltovski; • **103:** Luther C. Goldman/U.S. Fish and Wildlife Service; • **107:** rebeca; • **110:** Gilmartin Owen; • **111:** Jeff1980; • **112:** Imageafter; • **115:** cohdra; • **116:** (T) Modern Touch Photography; (B) natashaw; • **119:** (T) A_kartha; (B) beanworks; • **120:** (T-B) hamletnc; John Sarvis/U.S. Fish and Wildlife Service; Delboysafa; clarita; • **121:** (T) sveres; (C) Cx_ed; (B) kkiser; • **123:** Bain Collection, Prints & Photographs Division, Library of Congress, LC-USZ62-33430; • **124:** Imageafter; • **127:** xandert; • **128:** Holder; • **129:** (T) duboix; (B) United States Department of Agriculture;

132: (T) U.S. Navy photo by Mass Communication Specialist 1st Class Matthew D. Leistikow; (B) wadehjb; • **133:** Matthew Hull; • **135:** (T) ppdigital; (C) Paddleasia.com (B) Kevin Rosseel; • **136:** (T) Paddleasia.com; (C) shirleybnz; (B) ToOw1r3d; • **138:** Tony Nester; • **139:** CDC/Janice Haney Carr; • **140:** Imageafter; • **142:** Vestergaard Frandsen;

146: (TL) Tony Nester; (TR) Paddleasia.com; (BL) Tony Nester; • **147:** (All) Tony Nester;

148: Serious Fun; • **150:** (T) Jochanan Wahjudi; (B) MC1 Matthew D. Leistikow

151: (TL-BR) Homero chapa; alvimann; Imageafter; pickle; Matthew Hull;

152: (BL-TR) U.S. Navy photo by Mass Communication Specialist 1st Class Roger S. Duncan; Georgina DeBurca; seemann; bobby; Charl de Mille-Isles; US National Oceanic and Atmospheric Administration; • **153:** (T) U.S. Navy photo by Mass Communication Specialist 1st Class Roger S. Duncan; (c) Tony Nester; (B) Amejzing; • **154:** dantada; • **155:** (T) U.S. Navy photo by Mass Communication Specialist 1st Class Matthew D. Leistikow; (B) Chris Palmer; • **159:** (T) Prototype 7; (B) Cyndi Souza/U.S. Fish and Wildlife Service; • **160:** (BL-TR) mzacha; mzacha; U.S. Fish and Wildlife Service; blackbird; keithcr; • **161:** (TL-BR) jppi; heyjude; emenel; stock.xchg; cyanocorax; • **162:** (T) blackbird; (BL) Alfi007; (BR) Dan McKay; • **163:** (L) Kevin Rosseel; (C) acrylicartist; (R) MEJones; • **164:** (T,L-R) Donna A. Dewhurst; Steve Maslowski/U.S. Fish and Wildlife Service; jak; Steve Maslowski/U.S. Fish and Wildlife Service; (BL) DSchaeffer; (BR) gracey; • **165:** (TL-BR) Paddleasia.com; Katman1972; Floppy2009; Gittins; clconroy; Matthew Hull; Jimdaly98; Paddleasia.com; • **166:** (All) Paddleasia.com; • **167:** Tony Nester; • **169:** Tony Nester; • **171:** Barry Glickman; • **174:** (T) Ruth Lawson/Otago Polytechnic; (3BL) Tony Nester; • **175:** (T) schick; (B) mcleod; • **176:** (T) Tony Nester; (B) Don Mann; • **177:** Tony Nester; • **181:** johninportland; • **183:** NOAA; • **185:** (TL-BR) NOAA; NOAA; For Spacious Skies/John A. Day; NASA; NASA; • **186:** (TL-TR) Ralph F. Kresge/NOAA's National Weather Service (NWS) Collection; Ralph F.

Kresge/NOAA's National Weather Service (NWS) Collection; Dave; alvimann; (C) NOAA; (B) JASON Mission Center; • **187:** (T) Ralph F. Kresge/NOAA's National Weather Service (NWS) Collection; (B) NASA; • **188:** (T) bosela; (B) NASA; • **189:** (T) NASA; (C) NASA; (B) NOAA; • **190:** ostephy; • **193:** Harke; • **198:** (T) U.S. Navy photo by Seaman Stephen M. Fields; (B) U.S. Navy photo by Mass Communication Specialist 1st Class Matthew D. Leistikow; • **199:** (T) U.S. Navy photo by Mass Communication Specialist 1st Class Matthew D. Leistikow; (B) Earl53; • **200:** (T) rsvstks; (B) SPOT; • **201:** (T) TracMe; (C) ACR Cobham Beacon Solutions; (B) U.S. Navy photo by Mass Communication Specialist 1st Class Matthew D. Leistikow; • **202:** (T) cohdra; (B) iotdfi; • **203:** EmmiP; • **204:** (T) Irish_Eyes; (C) Livelovelaugh1; (B) Dimitri_c; • **205:** (All) NASA; • **206:** NASA; • **208:** (T) shuttermon; (C) Hainee; (B) hotblack; • **210:** wallyir; • **214:** (T) Berkeley; (B) Air Force; • **215:** (All) Dino Brown; • **216:** (T) Dr. Becky Abell; (B) James Gathany; • **217:** Serge8228; • **218:** (All) CDC; • **219:** CDC; • **222:** jeltovski; • **223:** Sarej; • **224:** Dino Brown; • **225:** Dino Brown; • **228:** jusben; • **229:** (All) Dino Brown; • **231:** (All) Dino Brown; • **232:** (All) Dino Brown; • **238:** (T) ASchaeffer; (C) Dylan Ellis; (B) Jelle80nl; • **239:** hconfer; • **240:** (T) Bain Collection, Prints & Photographs Division, Library of Congress, LC-DIG-ggbain-04779; (C) Bain Collection, Prints & Photographs Division, Library of Congress, LC-DIG-ggbain-31994; (B) aameris; • **241:** barry; • **243:** (3T) Courtesy of the Prints and Photographs Division, Library of Congress; (BR) clarita

**Use of released U.S. Navy imagery does not constitute product
or organizational endorsement of any kind by the U.S. Navy.**